Oxford Chemistry Series

General Editors
P. W. ATKINS   J. S. E. HOLKER   A. K. HOLLIDAY

Oxford Chemistry Series

1972

1. K. A. McLauchlan: *Magnetic resonance*
2. J. Robbins: *Ions in solution (2): an introduction to electrochemistry*
3. R. J. Puddephatt: *The periodic table of the elements*
4. R. A. Jackson: *Mechanism: an introduction to the study of organic reactions*

1973

5. D. Whittaker: *Stereochemistry and mechanism*
6. G. Hughes: *Radiation chemistry*
7. G. Pass: *Ions in solution (3): inorganic properties*
8. E. B. Smith: *Basic chemical thermodynamics*
9. C. A. Coulson: *The shape and structure of molecules*
10. J. Wormald: *Diffraction methods*
11. J. Shorter: *Correlation analysis in organic chemistry: an introduction to linear free-energy relationships*
12. E. S. Stern (ed): *The chemist in industry (1): fine chemicals for polymers*
13. A. Earnshaw and T. J. Harrington: *The chemistry of the transition elements*

1974

14. W. J. Albery: *Electrode kinetics*
16. W. S. Fyfe: *Geochemistry*
17. E. S. Stern (ed): *The chemist in industry (2): human health and plant protection*
18. G. C. Bond: *Heterogeneous catalysis: principles and applications*
19. R. P. H. Gasser and W. G. Richards: *Entropy and energy levels*
20. D. J. Spedding: *Air pollution*
21. P. W. Atkins: *Quanta: a handbook of concepts*
22. M. J. Pilling: *Reaction kinetics*

D. J. SPEDDING

Air pollution

Clarendon Press · Oxford · 1974

Oxford University Press, Ely House, London W.1

GLASGOW   NEW YORK   TORONTO   MELBOURNE   WELLINGTON
CAPE TOWN   IBADAN   NAIROBI   DAR ES SALAAM   LUSAKA   ADDIS ABABA
DELHI   BOMBAY   CALCUTTA   MADRAS   KARACHI   LAHORE   DACCA
KUALA LUMPUR   SINGAPORE   HONG KONG   TOKYO

CASEBOUND ISBN 0 19 855463 X
PAPERBACK ISBN 0 19 855464 8

© OXFORD UNIVERSITY PRESS 1974

PRINTED IN GREAT BRITAIN BY
J. W. ARROWSMITH LTD., BRISTOL, ENGLAND

Editor's Foreword

CHEMISTRY and pollution mean the same thing for many people, and the chemist is almost automatically regarded as a culpable polluter. After reading this book, the chemistry student should have no difficulty in correcting this view, and in shifting much of the culpability from man to nature. More importantly, he should see the relevance of much laboratory chemistry to the problems of atmospheric pollution, and recognize that the chemist has much to offer in the search for an atmosphere which is more congenial, in both the long and the short term, to man.

For more information about the atmosphere the reader should consult W. S. Fyfe: *Geochemistry* (OCS 16), while further discussion of motor-vehicle emissions and their control will be found in G. C. Bond: *Heterogeneous catalysis: principles and applications* (OCS 18).

A.K.H.

Acknowledgments

THE sources of data that have been published elsewhere are indicated in the text. Acknowledgements are gratefully made to the authors for permission to use the data, and to the following publishers for their permission to reproduce material from their publications.

Academic Press (*Photochemistry of air pollution*, by P. A. Leighton), Air Pollution Control Association (*Journal of the Air Pollution Control Association*), American Chemical Society (*Environmental Science and Technology*), American Geophysics Union (*Journal of Geophysics Research*), British Joint Corrosion Group (*British Corrosion Journal*), and the Clean Air Society of Australia and New Zealand (*Proceedings of the International Clean Air Congress*, Melbourne, 1972).

I wish also to express my sincere appreciation to Dr. P. Brimblecombe, Dr. J. R. Duncan and Mr. R. W. Meadows for their helpful comments, criticisms, and discussions during the preparation of the manuscript; to Professor Holliday for his very pertinent and helpful editorial advice; and to Miss S. Groves for her very competent typing of the manuscript.

Finally I wish to acknowledge the encouragement and enthusiasm of my wife, without which the manuscript would not yet be completed.

Contents

Introduction

AIR pollution studies are of a multi-disciplinary nature embodying subjects as widespread as sociology and physics, and law and botany. In this book an attempt has been made to cover the major aspects of air pollution emphasizing the discipline of chemistry. The material from other disciplines that appears in the text is not treated to the same depth as is that of chemistry. It is to be hoped that the material outside chemistry is sufficient to broaden the view of the reader specializing in chemistry and to stimulate him to read more widely on other aspects of air pollution.

Throughout the book it has been assumed that an air pollutant is some form of material added to the atmosphere as the result of the activity of man. Also an attempt has been made to show that, in almost all cases, materials regarded as atmospheric pollutants have a natural occurrence in the atmosphere. With this in mind it is hoped that the reader will emerge with a more balanced view of atmospheric pollution than that which prevails in the popular press.

Some attempt has also been made to reveal the poor state of knowledge of air pollution chemistry that currently prevails. Where conflicting theories exist, both have been given, so that the reader may assess which is the more likely. It is my hope that at least some of the readers of this book will be sufficiently stimulated by the breadth of research that remains to be done, that they will orient their work to the field of air pollution.

1. The atmosphere

THE earth's atmosphere is an envelope of gases extending to a height of about 2000 km. The density of these gases decreases with increasing altitude to such an extent that one half of the total mass of the atmosphere is found in the lower 5 km. Temperature also varies with altitude and this is used to divide the atmosphere into layers (Fig. 1.1). The properties of each of these layers have some relevance to air pollution chemistry although the properties of the troposphere are of the greatest significance since it contains the air that we breathe and the air in which all weather processes occur.

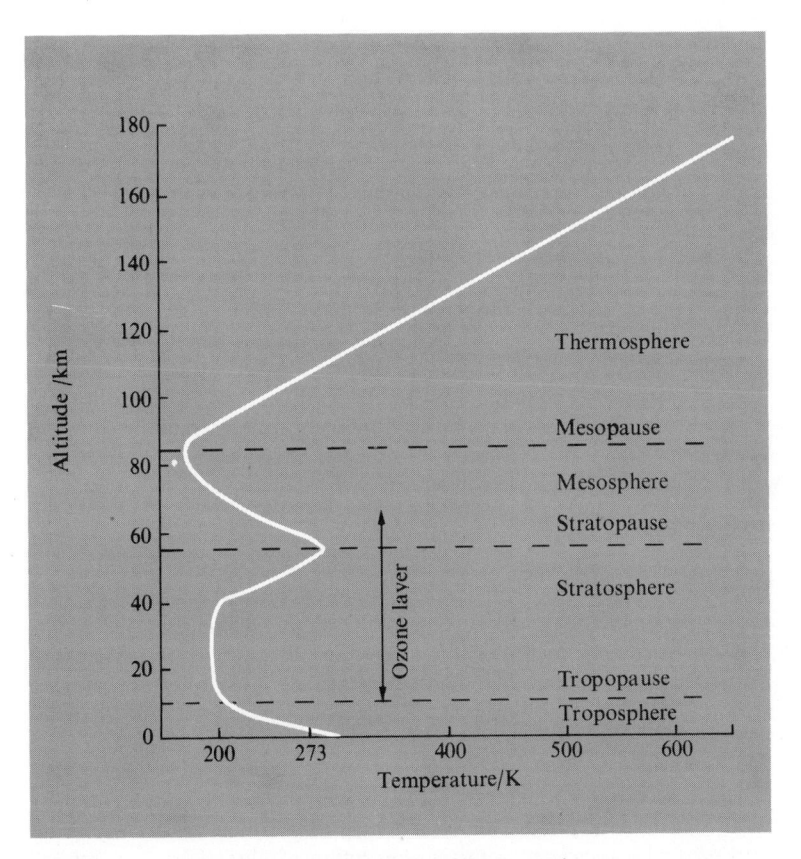

FIG. 1.1. Temperature profile of the atmosphere.

The troposphere

The troposphere is characterized by a steady decrease of temperature with altitude averaging 0·6°C per 100 m. It is maintained as a relatively distinct layer of the atmosphere by the cooler air of the stratosphere which lies above it. The portion of the atmosphere at which the negative temperature gradient of the troposphere changes to constant temperature is known as the tropopause. The tropopause is not continuous but generally has two and sometimes three separate levels of discontinuity at different latitudes in both hemispheres. The tropical tropopause has an altitude of about 16–18 km. There is a distinct break at 35–50° in both hemispheres between this layer and the polar tropopause which has an altitude of 9–10 km. A third tropopause, the mid-latitude tropopause, is found between the polar and tropical tropopauses and has an altitude of 10–12 km. The positions of the tropopauses oscillate in altitude and latitude from day to day and from season to season. Even during one 24 hour period the height of the tropopause above a given point may vary by several kilometers.

There is a relatively slow rate of exchange of material through the tropopause in either direction. Most of the exchange from the troposphere occurs at the tropical tropopause, while exchange to the troposphere occurs mostly in the mid-latitude tropopause. The lifetime for exchange from the stratosphere to the troposphere is of the order of months. This exchange rate is of obvious importance for pollutants injected into the stratosphere and which are destroyed in the troposphere or at the surface of the earth. Included in this category are some of the effluents of aircraft and the debris from atmospheric nuclear explosions.

Within the troposphere the mixing time of a given hemisphere is of the order of weeks, while complete exchange between hemispheres requires about a year. Much of the mass of gaseous air pollutants is emitted in the Northern Hemisphere, hence the slow inter-hemispheric mixing time is of importance when considering the global consequences of an atmospheric pollutant.

The stratosphere

The physical properties of the stratosphere are similar to those of the troposphere except that there is a reversal in the temperature gradient, with the temperature rising to 10–20°C at 60 km. Mixing within the layer is quite marked with strong horizontal air currents and considerable vertical mixing. Very little water vapour is found in the stratosphere thus processes associated with precipitation do not occur.

The mesosphere

The temperature of this layer falls again with increase in altitude reaching −70°C at the mesopause. The rise in temperature within the stratosphere is thought to be associated with the absorption of ultra-violet and infrared radiation from the sun by ozone (see Chapter 4). The ozone concentration in the mesosphere decreases rapidly with height, hence the temperature decrease is probably due to decreased absorption of solar radiation by ozone.

The thermosphere

This is also known as the ionosphere and is the highest layer yet recognized. It is characterized by a steady rise in temperature with altitude. The temperature at 200 km exceeds 500°C and at the upper boundary (700–800 km) it exceeds 1000°C. The temperature increase is related to the absorption of solar ultraviolet radiation by molecular oxygen and nitrogen.

The air in the thermosphere becomes ionized under the influence of solar radiation. The ionized particles are found in a number of sublayers in the thermosphere and are responsible for reflecting radio waves.

Composition of the atmosphere

Air is a rather stable mixture of gases whose relative proportions vary by no more than a few thousandths of one per cent near the surface (Table 1.1). There are some exceptions to this, the most important being water vapour which is confined almost exclusively to the troposphere by the processes of condensation and precipitation. Within the troposphere water vapour can reach up to 4 per cent by volume at some points, while at other points it can be almost totally absent.

TABLE 1.1

Composition of dry air at sea level

Component	Volume per cent
Nitrogen	78·084
Oxygen	20·946
Argon	0·934
Carbon dioxide†	0·321
Neon	0·001 82
Helium	0·000 52
Krypton	0·000 11
Xenon	0·000 008 7
Methane	0·000 125

† Value obtained in Antarctica in 1971.

Variable trace amounts of other gases are found in the atmosphere. These include various hydrocarbons, carbon monoxide, nitrogen oxides, hydrogen, ammonia, hydrogen peroxide, halogens, radon, sulphur dioxide, hydrogen sulphide, organic sulphides, and mercaptans.

Ozone shows a remarkable variation throughout the atmosphere, being found largely in the stratosphere. This is due to a combination of photochemical production and destruction reactions, together with the considerable reactivity of ozone toward other atmospheric components (see Chapter 4).

The relative proportions of the major permanent constituents remain almost unchanged up to at least 80 km. Above this point the production of atomic oxygen and atomic nitrogen becomes significant, thus changing the volume percentage composition of the two major atmospheric constituents.

It is of interest to note that all of the gases mentioned in this chapter can be produced by natural sources rather than by man's activity. These gases can hence be considered to be components of the unpolluted atmosphere.

2. Aerosols

AN aerosol in the atmosphere is defined as being a dispersion of solid or liquid matter in air. In this chapter it will be shown that, when the natural aerosol is supplemented by pollutant aerosol in the form of smoke, ash, acid mists, and soluble salts, the properties of the atmosphere may be adversely affected.

Processes of absorption and scattering of light by smoke and $(NH_4)_2SO_4$ aerosols greatly affect visibility. The absorption of gases by the solid aerosol provides a means of transmitting soluble toxic gases deeper into human lungs than would be the case for the gas alone. The deposition of solid aerosols, especially soot, on exterior surfaces of buildings is a very obvious, and expensive, form of pollution by solid aerosols. The deposition of acid mists, especially H_2SO_4 mist leads to accelerated corrosion of metals while mists of carcinogenic hydrocarbons from combustion processes have obvious effects on human health.

Particle size

The particle size of the aerosol is controlled by physical processes. The upper limit is under the control of gravitational forces while the lower limit is controlled by coagulation processes. Table 2.1 lists some of the nomenclature which apply to the atmospheric aerosol. It can be seen that particle size is of basic importance. Electron microscopy shows that solid particles in the atmosphere vary markedly in configuration, ranging from almost spherical fumes to dusts of very irregular shape. The usual method of expressing particle size is the Stokes radius which is defined as the radius of a sphere having the same falling velocity as the particle, and a density equal to that of

TABLE 2.1

Terms used to describe the atmospheric aerosol

Aitken particles	Particles of less than 0.1 μm radius
Large particles	Particles of radii in the range 0.1 to 1 μm.
Giant particles	Particles of radii greater than 1 μm
Dust	Solid particles broken down from solid material and dispersed by air currents.
Fume	Solid or liquid particles formed by condensation in the vapour phase
Smoke	A fume formed as the result of a combustion process.

the material in the particle. The Stokes radius of an isolated fume particle is almost identical with its geometrical radius but the Stokes radius of a dust particle formed from the coagulation of several other particles may be very much less than the measured 'radius'. Care is thus necessary in interpreting possible physical effects due to particle size, especially optical effects.

Size distribution

It is valuable to be able to express the number of particles of a given size for all the particle sizes in the atmospheric aerosol. Both particle number and particle size range over several orders of magnitude so a logarithmic representation is used (eqn 2.1).

$$n(r) = \frac{dN}{d(\log r)} \, cm^{-3} \tag{2.1}$$

where N is the total number of aerosol particles of radius smaller than r and $n(r)$ is the number of particles of radius between r and $r + \delta r$. Fig. 2.1 shows the usual graphical representation of this function. The graph is basically a curve through a histogram, hence the total number of particles of radius between r and $r + \delta r$, is represented by the area under the curve between these radius limits.

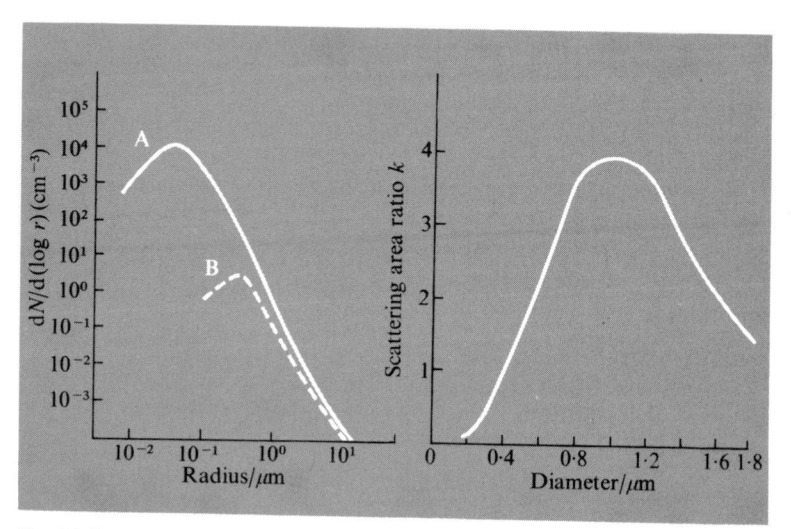

FIG. 2.1. Log number–radius distribution for a continental aerosol (A) and a maritime aerosol (B).

FIG. 2.2. Scattering area ratio as a function of particle diameter for spherical particles at wavelength 524 nm.

The log radius–number distribution for the aerosol found over large land masses (continental aerosol) shows a maximum at about 0·03 μm radius, i.e. most particles in the continental aerosol are Aitken particles. On the other hand it can be shown that the Aitken particles contribute no more than 20 per cent of the total mass of the continental aerosol. The remainder of the mass is approximately evenly divided between the large and giant particles. The aerosol found over the ocean (maritime aerosol) has a lower concentration of particles and a number maximum at greater than 0·1 μm radius in the log radius–number distribution. As sea salt forms almost all of the mass of the maritime aerosol its average particle radius varies with humidity, which controls the condensation of water vapour on the salt crystal. There is almost an order of magnitude difference between the radius of a dry sea salt crystal at low humidity and the radius of the droplet formed on this crystal at 99 per cent relative humidity.

A simplified explanation for the log radius–number distribution of the atmospheric aerosol is that it represents a balance between coagulation and sedimentation processes. The very small particles show a strong tendency to coagulate, forming particles of higher radius, while the very heavy particles are few in number due to their high sedimentation rate.

Sedimentation

When a particle falls through air a frictional drag due to the viscosity of the air is exerted on the moving particle. In order to maintain a uniform velocity, a constant force must be applied to overcome the viscous drag of the air. Stokes Law (eqn 2.2) expresses the magnitude of this force.

$$f = 6\pi r \eta u \tag{2.2}$$

where r is the radius of the small sphere, u is the velocity of the sphere and η is the coefficient of viscosity of the air. If the particle is falling under the influence of gravity the downward force F_1 is given by eqn (2.3).

$$F_1 = 4/3\pi r^3 (p - p')g \tag{2.3}$$

where p is the density of the particle, p' is the density of air and g is the acceleration due to gravity. At a certain point in the fall of the particle, constant velocity will be reached, i.e. $f = F_1$ thus:

$$4/3\pi r^3 (p - p')g = 6\pi r \eta u \tag{2.4}$$

or

$$u = \frac{2gr^2(p - p')}{9\eta} \tag{2.5}$$

When the particle radius approaches the mean free path of air the viscous drag of the air may be regarded as discontinuous, and some of the particles may 'slip' between air molecules. A correction factor may be applied to eqn (2.5) to account for this.

Eqn (2.5) shows that the terminal velocity u of an aerosol depends upon the square of its radius. The range of terminal velocities for the atmospheric aerosol is thus very wide e.g. at 101·3 kPa pressure and 20°C a spherical particle of density $1\,\mathrm{g\,cm^{-3}}$ and radius 0·1 μm has a terminal velocity of $9 \times 10^{-5}\,\mathrm{cm\,s^{-1}}$ while a similar particle of radius 20 μm has a terminal velocity of 1·2 $\mathrm{cm\,s^{-1}}$.

Sedimentation controls the upper limit of atmospheric aerosol size. It is of interest to note that natural particles, such as pollens and loess (wind-blown soil) have a remarkably uniform radius of about 10 μm, which suggests that particles with terminal velocities less than those of pollens and loess (about $0.3\,\mathrm{cm\,s^{-1}}$) remain airborne for some time.

Coagulation

Coagulation occurs when particles, rather than bouncing apart on contact, accrete or coalesce. The process is a continuous one, so that the number of particles decreases with time, while the size of the particles increases with time. Smoluchowski has derived an equation (eqn 2.6) to describe the change in concentration with time, of the number of homogeneous particles in a given volume.

$$-\frac{dn_t}{dt} = Kn_t^2 \qquad (2.6)$$

where n_t is the number of particles present at time t and K is a constant that depends upon particle size and diffusivity and a number of physical properties of the air. The smaller a particle the higher its diffusivity, hence very small particles have a high rate of coagulation. Particles of less than 0·01 μm radius have diffusion coefficients greater than $5 \times 10^{-4}\,\mathrm{cm^2\,s^{-1}}$ and are rapidly coagulated. Particles larger than 0·3 μm radius have diffusion coefficients of less than $2 \times 10^{-6}\,\mathrm{cm^2\,s^{-1}}$ which is negligibly small and thus act as acceptors on to which the rapidly diffusing smaller particles coagulate. The larger particles so formed are removed by sedimentation.

Condensation of water

The normal particle present in the atmospheric aerosol is regarded as being a mixture of soluble and insoluble materials. The soluble fraction is very important in the formation of droplets in the atmosphere. When the relative humidity of the air above a dry soluble salt exceeds the relative humidity above a saturated solution of that salt, water is condensed on to the particle containing the salt. The dry particle thus becomes a saturated salt solution containing some insoluble material. As the relative humidity in the air increases, so the radius of the droplet increases and the solution becomes less saturated. As we have seen in the previous section, the larger the particle the greater the chance of deposition by sedimentation. The probability of rainfall at high relative humidity is thus related to the condensation of water vapour on the atmospheric aerosol.

In the maritime aerosol, NaCl is an important soluble salt. At a relative humidity of 75 per cent at 20°C a dry crystal of NaCl becomes a saturated solution. In the continental aerosol, $(NH_4)_2SO_4$ is an important soluble salt. At a relative humidity of 81 per cent at 20°C a dry crystal of $(NH_4)_2SO_4$ becomes a saturated solution. The predominant salt in the atmospheric aerosol therefore has an influence on the formation of droplets at a given relative humidity.

It can be shown that the large and giant particles are very much more important in droplet formation than are the Aitken particles. The reason for this can be seen from eqn (2.7).

$$P/P_0 = 1 + \frac{c_1}{r} - \frac{c_2}{(r^3 - r_0^3)} \qquad (2.7)$$

where P/P_0 is the relative humidity in equilibrium with the droplet, r is the radius of the droplet, r_0 is the radius of the insoluble matter and c_1 and c_2 are constants. The term c_1/r accounts for the increase in relative humidity due to the Thomson effect of curvature. For radii above 0.1 μm this is smaller than 1 per cent relative humidity but it increases with decreasing radius reaching about 500 per cent relative humidity at a radius of 10^{-3} μm. Above 0.1 μm, therefore, the relative humidity is controlled by the last term in eqn (2.7). This term expresses the depression in water vapour pressure due to the concentration of salt in solution, i.e. Raoult's Law.

Most of the mass of atmospheric aerosol resides in the large and giant particles. These particles are also the most important in condensing water vapour and thus producing rain. Much of the mass of aerosol in the atmosphere is thus removed by washout in rain.

Effects on visibility

The most obvious effect of atmospheric pollution is a reduction in visibility. When visibility is reported in meteorological data, it refers to a measurement of the greatest distance at which a dark object of reasonable size may be seen against the horizon sky. The ability to see such an object depends upon the transmissions of light through the atmosphere, and on the contrast of the object to the background. Both of these factors are influenced by absorption and scattering processes in the atmosphere.

If a parallel beam of light is transmitted through a uniform atmosphere its intensity I falls exponentially with distance x (see eqn 2.8).

$$I = I_0 e^{-\sigma x} \qquad (2.8)$$

Here I_0 is the original intensity and σ is the extinction coefficient. The extinction coefficient may be expressed as the sum of scattering effects (b) and absorption (k) as in eqn (2.9).

$$\sigma = b + k \tag{2.9}$$

Atmospheric gases may have an effect on the extinction coefficient by both scattering and absorption. Scattering occurs mainly as Rayleigh scattering, where the frequency of the incident light is unchanged in the scattering process. This occurs with gas molecules and particles of size much smaller than the wavelength of the incident light i.e. less than 0.1 μm radius. Aitken particles are thus of importance in this phenomenon. The contribution of Rayleigh scattering in the reduction of atmospheric visibility is, however, very small compared with other effects.

The gases normally present in the atmosphere do not absorb visible light and, of the pollutant gases, only NO_2 is present in sufficient concentration to have any significant effect. It can be calculated that, in an atmosphere with a visual range of 16 km in the absence of NO_2, a concentration of 0.5 mg kg^{-1} NO_2 would produce a significant discolouration of the atmosphere. The discolouration is yellowish-brown, as NO_2 absorbs significantly in the blue–green portion of the visible spectrum.

The scattering caused by the atmospheric aerosol is due to particles of a size comparable to the wavelength of visible light (0.4–0.8 μm) i.e. large particles. This type of scattering is known as Mie scattering, and the scattering coefficient b in this case may be obtained from eqn (2.10).

$$b = N K \pi r^2 \tag{2.10}$$

where N is the number of particles of radius r, and K is the scattering area ratio for these particles. This ratio is dependent upon the particle radius and refractive index, and on the wavelength of the incident light (see Fig. 2.2, see p. 8).

The absorption of light by the atmospheric aerosol appears to have an effect comparable to the scattering of light by Mie scattering. The absorption effect is, of course, related to the colour of the aerosol particles.

The preceding discussion relates to the reduction in light intensity due to the components of the atmosphere. The meteorological definition of visibility relates to the contrast of an object relative to the background. The contrast is affected by the processes influencing light transmission and it can be shown that the apparent contrast C at a distance x can be determined from eqn (2.11).

$$C = C_0 \, e^{-\sigma x} \tag{2.11}$$

where C_0 is the actual contrast of the object and σ is the extinction coefficient. In the definition of visibility the target is assumed to be black, i.e. $C_0 = -1$, hence eqn (8.11) may be written (eqn 2.12),

$$-C = e^{-\sigma x}. \tag{2.12}$$

The eye is used as the sensor in the determination of visibility and it is assumed to have the ability to determine a contrast down to a limit of 2 per cent i.e. $C = 0.02$. If this value is applied to eqn (2.12) the distance x becomes visual range V, by the definition of visual range (eqn 2.13),

$$e^{-\sigma x} = 0.02. \tag{2.13}$$

From this V is found to have the value given in eqn (2.14),

$$V = \frac{3.9}{\sigma}. \tag{2.14}$$

An example of the use of this concept is illustrated in Table 2.2. At a constant ammonium ion concentration the value of σ, measured with an integrating nephelometer, falls with decreasing humidity. The visual range thus increases with decreasing humidity. We have earlier seen that the relative humidity of the air above a soluble salt influences the radius of the droplet formed by water condensing on the salt (or solution). It can also be seen (e.g. eqn 2.10), that the radius of a particle influences the value of σ. The data in Table 2.2 thus illustrate the importance of soluble salts in the atmospheric aerosol, in determining the visibility of the atmosphere (see also the section on ammonium sulphate, Chapter 8).

TABLE 2.2

Relationship between σ and particulate NH_4^+ concentrations at Stockton, England

Ammonium concentration ($\mu g\ m^{-3}$)	Relative humidity (%)	σ (km^{-1})	V (km)
20	96–97	3.7	1.1
20	89–91	2.2	1.8
20	84–86	1.2	3.3

Chemical composition of the tropospheric aerosol

The chemical nature of the tropospheric aerosol has a significant effect on the properties of the troposphere, the most important effects being in water condensation and visibility. The tropospheric aerosol has two major sources— the ocean and the land. The maritime aerosol, to a reasonable approximation, is of similar composition to the ocean, while the continental aerosol contains materials dispersed from the surface of the earth from both natural and man-made sources. Any given aerosol is a mixture of particles from both maritime and continental sources, with the maritime aerosol predominating in mid-ocean, and the continental aerosol predominating in the centre of land masses.

Man is concerned with the aerosol deposited over the land therefore the discussion following will be largely oriented to the chemical composition of that aerosol. Attempts to determine the composition of the Aitken particles in the tropospheric aerosol, have to a large extent, been unsuccessful. The large and giant particles have, however, been analyzed. Table 2.3 sets out average results for the distribution of some soluble salts between large and giant particles.

TABLE 2.3

Distribution of soluble salts between large and giant particles in the atmosphere $(\mu g\ m^{-3})$

	Large		Giant	
	Junge	Novakov *et al.*	Junge	Novakov *et al.*
Cl^-	0·03	—	1·2	—
Na^+	0·02	—	1·2	—
SO_3^{2-}	—	2·86	—	0·42
SO_4^{2-}	4·6	1·43	1·2	0·65
NO_3^-	0·06	0	0·7	0·14
NH_4^+	0·8	0·5	0·2	0·14
Amino N	—	1·6	—	0·25
Pyridino N	—	2·0	—	0·41

The data of Junge show that NH_4^+ and SO_4^{2-} predominate in the large particles and that they exist in a mole ratio that suggests that $(NH_4)_2SO_4$ is the salt present. This has been confirmed by electron microscopy studies, and it is now known that $(NH_4)_2SO_4$ is widely distributed in the atmosphere. The importance of atmospheric $(NH_4)_2SO_4$ will be discussed in Chapter 8.

The data of Novakov *et al.* are for an aerosol from an area subject to photochemical smog episodes (Pasadena, U.S.A.). The distribution of NH_4^+ and $(SO_4^{2-} + SO_3^{2-})$ between the large and giant particles is similar to that found by Junge but the mole ratios show an excess of $(SO_4^{2-} + SO_3^{2-})$ over NH_4^+. In this aerosol a significant portion of the sulphur anions existed as salts other than ammonium salts, or as H_2SO_4 and H_2SO_3.

The Na^+ and Cl^- ions are found largely in the giant particle size range. It has been found that particles of maritime origin tend to be in this size range (see size distribution section) and elements and compounds of oceanic origin are thus concentrated in this size range. On the other hand, particles of continental origin tend to be in the large particle size range. Particles containing amino-nitrogen and pyridino-nitrogen are found mostly in the large particle size range. It is likely that these are emitted directly to the atmosphere from motor vehicles, as petrol is known to contain compounds with these functional groups.

TABLE 2.4

Concentration of some elements and salts from districts of different geography

	Windermere (μg kg^{-1})	Stockton (μg m^{-3})	Pasadena (μg m^{-3})
Total	20	—	101·5
Al	0·26	0·8	0·8
Br	0·027	0·09	0·6
Ca	0·52	1·3	0·99
Cl	1·75	2·8	0·07
Cu	0·026	—	0·03
Cr	0·002	0·008	—
Fe	0·23	1·7	3·2
I	0·002	—	0·006
K	—	0·3	0·32
Mg	—	0·5	1·1
Mn	0·01	0·1	0·03
Na	2·3	0·8	1·0
Pb	0·09	0·4	3·3
V	0·008	0·02	0·01
Zn	0·08	—	0·18
NO$_3^-$	—	2·2	9·5
NH$_4^+$	—	4·4	0·6
SO$_4^{2-}$	—	11·4	12·1

Table 2.4 lists average concentrations of elements found at three sites of distinctly different geography. Lake Windermere is a rural site in Great Britain, 30 km from the nearest heavy industry; Stockton is an industrial district in Great Britain, while Pasadena is a district in the United States known for a high incidence of photochemical smog. Motor vehicle activity at the three sites is reflected especially in the concentrations of Pb, Br, and NO$_3^-$. Both Pb and Br are components of 'anti-knock' fluid in petrol and appear in highest concentration in Pasadena which has a higher rate of automobile activity than has Stockton. The low values obtained for Pb and Br at rural Windermere reinforce the view that these elements are of automobile origin.

The elements Na and Cl are regarded as being good indicators of the presence of sea-salt aerosol. In sea water, the ratio of Cl:Na is about 1·8. At Windermere the ratio is slightly lower than this, but sufficiently close to suggest the presence of a considerable amount of maritime aerosol. At Stockton the ratio is very high and probably reflects the presence of industrial sources of Cl. In the Pasadena aerosol the ratio is very low, probably because of reactions of Cl with photochemical smog components. It is thus obvious that the reactivity of elements, and possible anthropogenic sources, must be considered in attempting to determine their origin in the atmospheric aerosol.

It is possible to show an anthropogenic contribution to the atmospheric aerosol by selecting elements that are not likely to be of natural origin. In Table 2.4 an example of such an element is vanadium. It can be seen that in industrialized Stockton the concentration of this element is considerably higher than in Windermere or Pasadena.

The total concentration of particulate matter shown in Table 2.4 is three or four times greater than the sum of the concentrations of the elements listed. Much of this deficit appears as non-carbonate carbon, which can form up to 45 per cent of the total particulate concentration. Carbonate carbon usually forms less than 2 per cent of the total particulate concentration and appears mainly in the giant particle size range. 90 per cent of the non-carbonate carbon is found in the large particle size range.

A wide range of organic compounds has been found in the atmospheric aerosol. These include all the straight-chain alkanes from C_{18} to C_{34}, at least thirty polycyclic hydrocarbons, and many heterocyclic compounds. The aromatic polycyclic hydrocarbons are of particular interest because a number of them are known carcinogens. The best-known of this group are the benzpyrenes which are found in the atmosphere at a concentration of about 5 ng m^{-3}. They are obtained from the high temperature combustion or pyrolysis of carbonaceous materials, particularly coal tar.

Chemical reactions of atmospheric aerosols

It is to be expected that the most important reactions of atmospheric aerosols relate to their ability to absorb gases and/or catalyze gaseous reactions. When an aerosol is at equilibrium with the gases and vapours in the atmosphere it has a specific surface area of about 2–2.5 m^2 g^{-1}. It can be shown that this is approximately one half of its total specific surface area, the other half being occupied by atmospheric gases and vapours. The high surface area is accounted for by the presence of a large number of micropores most of which have a radius of less than 10 nm. These pores absorb and condense atmospheric gases including CO_2 (150–$300\,\mu$g g^{-1}) CO (10–$30\,\mu$g g^{-1}), CH_4 (15–$60\,\mu$g g^{-1}) and NH_3 (30–$100\,\mu$g g^{-1}). It is of interest to note that SO_2 in gaseous form has not been identified on an atmospheric aerosol despite the fact that most urban aerosols contain 2–4 per cent by weight of sulphur. This suggests that adsorbed SO_2 is rapidly oxidized to sulphate on aerosol particles. The presence of acid particles in the atmosphere is well known in areas of high SO_2 concentration. The acidity of these particles is attributed to H_2SO_4 formed by SO_2 oxidation. Of course, if the particle contains alkali metals or alkaline earth metals then the acidity is neutralized and sulphate salts are formed.

Smoke and ash

These are aerosol particles arising from combustion processes which may be of natural or man-made origin. Natural combustion processes include grass, brush, and forest fires as well as volcanic activity. An average grass fire extending over one acre will produce about 10^{22} fine particles which, if uniformly distributed throughout a column of air of height 3 km and cross-sectional area 4000 m^2, would give a concentration of about 10^9 particles per cm^3. The average diameter of a smoke particle is about 0·075 μm, i.e. most are Aitken particles. Much of the smoke is made up of carbonaceous compounds, particularly tarry hydrocarbons and resins. The very small size of smoke particles enables them to penetrate buildings in the same manner as gases. Unlike gases, the smoke particles have a high 'sticking' power and are found deposited on surfaces where air turbulence is high e.g. edges of window and door frames.

The solid material set free when a fuel is completely oxidized is known as ash. The size of the particles emitted from an industrial furnace depends upon the lowest velocity of the combustion gases in the chimney. We have seen, in the discussion on sedimentation, that the terminal velocity of a particle increases with size. In order that a particle is emitted from a chimney the flue gas velocity must exceed the terminal velocity. Particles of size such that the terminal velocity is not exceeded remain in the combustion system as ash. A typical flue-gas velocity in an industrial chimney is 12 ms^{-1}, which is sufficient to carry particles of $\leqslant 200\ \mu m$ diameter out of the chimney. The larger particles sediment close to the chimney as the flue-gas velocity decreases, while the smaller particles travel further from the chimney before sedimenting. The flue-gas velocity in a domestic open-fire chimney is rarely greater than 1·5 ms^{-1}. In this case the maximum size of ash particle leaving the chimney is 75 μm in diameter.

Damaging effects of aerosols

Some of the atmospheric aerosol may have a damaging effect on human health because of its chemical nature. The presence of carcinogenic hydrocarbons has already been mentioned. Elements such as Pb and As may exhibit toxic effects but only at quite high concentrations which are not normally experienced in the atmosphere. The aqueous aerosol can also be damaging to human health because of its chemical nature. Sulphuric acid mists formed during the atmospheric oxidation of SO_2 are known to cause respiratory damage to laboratory animals at concentrations of 2·5 $mg\,kg^{-1}$. Mists of oils from industrial processes and from motor vehicles do not specifically damage human health at concentrations found in polluted atmospheres.

The physical properties of the atmospheric aerosol affect human health, either by allowing penetration of the lung and causing irritation to the internal membranes, or by transporting adsorbed toxic gases and vapours deeper into the lung than they would normally travel. Many gases are soluble in the layer of water covering the mucous membranes of the respiratory tract. Some of the very soluble gases are completely removed on these surfaces before reaching the lungs. However, very small particles are transported deep into the lungs and the gases adsorbed on them may be released within the lungs (see Chapter 5).

Materials in our environment may be damaged by the larger solid particulate matter by mechanical abrasion. More important however, is the deposition of the atmospheric aerosol on materials, especially buildings. Little damage is caused to the building surfaces, but the effect is unsightly and expensive to remove. Accelerated attack by corrosive gases especially SO_2 is noted in the area about deposited particulate matter on metal surfaces (see Chapter 5). This may be due to the formation of distinct electrochemical cells on the particulate material and the metal, or it may simply be due to the ability of the particulate material to concentrate the corrosive gas at a point by absorption.

3. Carbon dioxide and water vapour

THE combustion of fossil fuels to provide a source of energy is the major means by which man pollutes the atmosphere. Almost all of the fossil fuels (coal, oil, natural gas, etc.) are carbonaceous or hydrocarbon in nature. Optimum usage of these fuels requires complete oxidation in the combustion process. Carbon dioxide and water vapour are thus, in terms of total mass emitted, the major gaseous pollutants of the atmosphere. Both gases have similar effects on climate but their atmospheric reactions are so different that they will be discussed separately.

The effect of pollutant water vapour will be shown to be very small, while the effect of carbon dioxide will be seen to be possibly of considerable consequence. The concentration of carbon dioxide in the atmosphere is still showing a steady increase. It is theoretically possible to show that this may result in an increase in the temperature of the earth—the greenhouse effect. The extent to which the greenhouse effect will change the pattern of the world's climate depends upon the numerical parameters used in a theoretical model. Parameters that predict rapid disastrous climatic effects have been used while some other parameters that predict limited climatic effects have also been used. It is probably wise to err on the side of caution and seek to limit the emission of carbon dioxide to the atmosphere and thus allow the total carbon cycle to come to equilibrium through the vast reservoirs of the deep ocean and sediments.

Water vapour

Almost all of the water vapour in the atmosphere is confined to the troposphere. The total mass of atmospheric water vapour is equivalent to a precipitation of 2·5 cm of rain over the whole of the earth's surface. The average rainfall over the earth is about 90 cm per year hence there are about thirty-six evaporation–precipitation cycles per annum. This means that the average residence time for a water molecule in the atmosphere is about ten days. This is of particular importance in air chemistry as many trace substances are removed from the troposphere by water precipitation.

Substances may be removed from the atmosphere by water precipitation in one of two ways—rainout or washout. Rainout occurs within clouds where the most important process is condensation of water vapour on the solid aerosol (see Chapter 2). Much of the total mass of the natural aerosol is removed from the atmosphere when this water precipitates. Washout occurs below the clouds and is a very efficient process for the removal of large solid aerosols. After an appreciable amount of rain, however, almost all of the solid aerosol is removed from below the cloudbase. The often dramatic improvement in visibility following a shower of rain can be explained by the washout of much of the light-scattering aerosol that makes such a large contribution to haze.

Many of the soluble gases e.g. CO_2, NH_3, SO_2, NO_2 are also removed from the atmosphere by rainout and washout. As an example, the removal of SO_2 as sulphate may be quoted. Of the total sulphate concentration in rainwater it has been calculated that 70 per cent is derived from SO_2 collected during washout and only 5 per cent from SO_2 collected during rainout.

Liquid water in the atmosphere has an average residence time of about eleven hours, which is longer than the average lifetime of clouds. It can be concluded then, that most clouds evaporate. Solids dissolved or suspended in the liquid water of clouds become solid aerosol particles when clouds evaporate, so that the presence of clouds does not necessarily indicate a rapid cleansing of the atmosphere.

The heat balance of the earth is influenced by atmospheric water, both in the liquid and in the gaseous state. Only 47 per cent of the total solar radiation reaching the thermosphere is received by the surface of the earth. Liquid water in the form of clouds reflects about 20 per cent of the incoming radiation. The surface of the earth reflects about 14 per cent while the remainder is absorbed by ozone in the stratosphere, and carbon dioxide and water vapour in the troposphere.

In Fig. 3.1 it can be seen that the black-body spectrum for solar radiation is largely in the ultraviolet–visible region of the spectrum. Also shown are the absorption spectra of O_3, O_2, NO_2, H_2O and CO_2 in the same wavelength region. Ozone shows a strong absorption in the ultraviolet region and it is this that shields the earth's surface from much of the ultraviolet radiation of the sun. Little absorption can be found in the visible region while in the infrared the predominant absorption is by water vapour, which absorbs 10–20 per cent of solar radiation.

Terrestrial radiation from earth to space is due to black-body radiation at the temperature of the earth's surface i.e. about 10°C. This radiation spectrum is also shown in Fig. 3.1. Terrestrial radiation is infrared radiation and is strongly absorbed by both CO_2 and water vapour. Only radiation of 7000 nm to 14 000 nm can escape directly to space through a 'window' of low absorption in the water-vapour spectrum. Terrestrial radiation absorbed by CO_2 and water vapour increases the temperature of these gases. This energy is eventually re-emitted with a considerable portion being directed back to the earth's surface. Radiation reflected in this manner serves to keep the temperature at the surface of the earth at a higher level than would be found if the gases were absent. This effect is called the 'Greenhouse Effect', because of the obvious analogy with the warming effect of a greenhouse.

Any increase in the atmospheric water vapour concentration will necessarily increase the temperature at the earth's surface because of the greenhouse

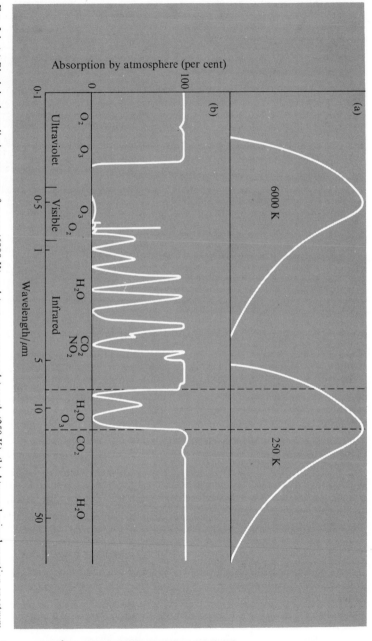

Fig. 3.1. (a) Black-body radiation spectra for sun (6000 K) and (not on same scale) earth (250 K). (b) Atmospheric absorption spectrum produced by principal absorbing gases (after Sawyer (1972)).

effect. Man's activities, however, produce only a very small portion of the total water in the water cycle. The atmospheric water vapour content is thus not measurably affected, leaving us with the conclusion that water vapour is a very innocuous pollutant gas.

Carbon dioxide

All important processes involving atmospheric CO_2 occur at the earth's surfaces. These processes may be summarized in the carbon cycle illustrated in Fig. 3.2. The atmosphere acts as a passive buffer reservoir with a large but limited capacity. The ocean is a reservoir sixty times as large as the atmosphere and is made up of two distinct layers. The surface layer of the ocean is about 100 m deep and is in a much more rapid equilibrium with the atmosphere than the deep layer which forms the bulk of the ocean. By far the largest portion of carbon in the carbon cycle is found in the marine and terrestrial sediments. The turnover of carbon within this reservoir is very slow indeed e.g. marine carbon is estimated to turn over once in 100 000 years. Man interferes with this cycle by using carbonaceous fossil fuels as an energy source and thus increases the turnover of sedimentary carbon by the production of CO_2 which is emitted into the atmosphere. The effect of this activity will be discussed later in this chapter.

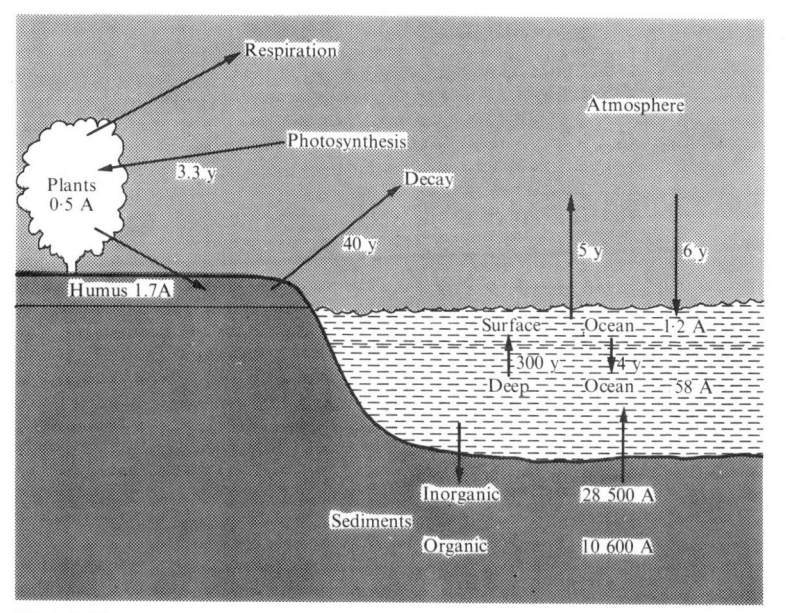

FIG 3.2. The carbon cycle. The times are residence times for CO_2 with respect to the indicated reservoir. $A = 2.5 \times 10^{18}$ g of CO_2 = atmospheric CO_2 reservoir.

In general there is no real CO_2 balance at any given point on the earth's surface. Plant life, in particular, affects the CO_2 concentration near the earth's surface by photosynthesis and respiration. Photosynthetic assimilation of CO_2 ceases at night leading to higher CO_2 concentrations at night. During the day the reverse effect may be found e.g. CO_2 concentrations below potato leaves in the field can be up to 30 per cent higher than those above the leaves. Such fluctuations may still be found 1 and 2 km into the troposphere.

One of the important steps in the carbon cycle is the exchange of CO_2 between the atmosphere and the ocean surface layer. Gas exchange at an aqueous surface is of considerable importance in atmospheric chemistry so it will be dealt with in some detail here.

At equilibrium, the mass of a gas that will dissolve in an aqueous solution is directly related to the partial pressure of that gas in the gas phase. The higher the gas-phase partial pressure, the higher the partial pressure of the gas in solution. This relationship is summarized in Henry's Law (eqn 3.1);

$$P = HX,\qquad(3.1)$$

where P = partial pressure of the gas expressed as a mole fraction,

X = concentration of the gas in solution expressed as a mole fraction,

H = Henry's Law constant.

This law applies to an equilibrium situation, hence over a long time scale the amount of a given gas in solution in natural water must be in diffusion equilibrium with its partial pressure in the atmosphere. However, at any given instant, the partial pressure of the gas in the water may be above or below that in the overlying atmosphere. There will thus be a net flux of the gas one way or the other across the interface tending to bring the system back to equilibrium. To a reasonable approximation the atmosphere can be considered to be of constant composition and nearly constant pressure. The direction of movement of a gas across the air–water interface therefore depends mostly upon factors that change its partial pressure in the surface waters. These factors include seasonal water temperature variations (the solubility of CO_2 increases with decreasing temperature), and photosynthesis and respiration of aquatic plants.

When such a partial pressure difference exists the kinetics of the exchange of the gas between the two phases must be considered. It is usual to discuss these kinetics in terms of an 'exchange constant' k which is defined as follows:

$$\frac{dq}{dt} = kA(X_1 - X_2)\qquad(3.2)$$

where dq/dt = rate of change of mass q of CO_2 with respect to time t,

A = area of the interface across which exchange takes place.

X_1 and X_2 represent the CO_2 concentrations in mass per unit volume in each phase.

k thus has the dimension of length time^{-1}.

With a natural water surface there is approximately a constant surface area so that the rate of net mass exchange of gas depends upon the partial pressure difference between the two phases and upon k. The physical significance of k may be described by the liquid-film model. In this model it is assumed that in both the liquid and gas phases there is a thin film of fluid where there is locally laminar flow while the bulk of the fluid is in turbulent flow (Fig. 3.3).

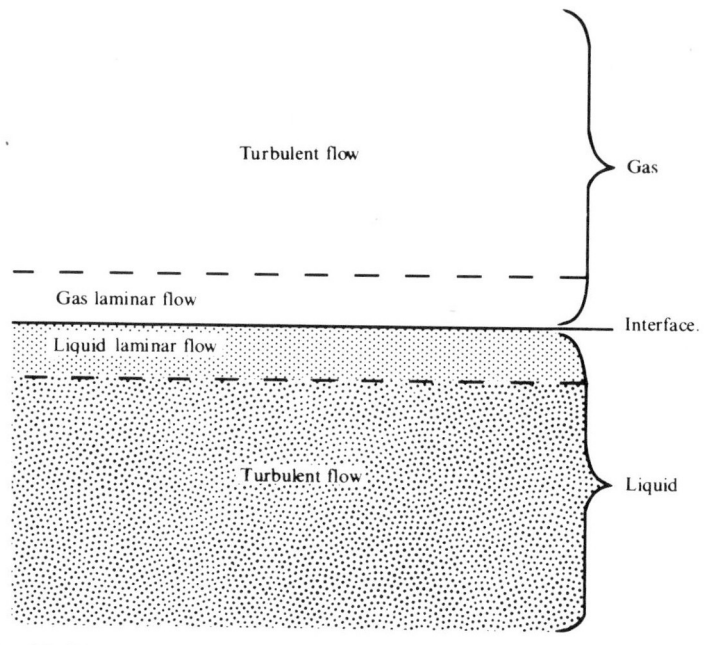

FIG. 3.3. Diagram to illustrate the laminar-layer model as applied to a gas–liquid interface.

Gas exchange between the two phases is thought to be controlled by the molecular diffusion of the gas through both laminar layers. In the case of CO_2 the diffusion in the liquid phase is about ten thousand times slower than in the gas phase and is thus rate-determining. It is also possible to relate k for CO_2 with the molecular diffusivity D of CO_2 in the liquid and the thickness of the liquid laminar layer δ:

$$k = \frac{D}{\delta}. \tag{3.3}$$

We can see now that the rate of exchange of atmospheric CO_2 with the ocean surface depends upon the CO_2 partial pressure difference between the phases, and on the thickness of the laminar layer of the ocean. This latter varies from 10–200 μm and is determined by the wind velocity immediately above the surface. The shear stress of the wind tends to tear the laminar layer film from the bulk of the water, thus high wind velocities lead to low values of δ. Because ocean conditions vary from glassy calm where δ is at its maximum value, to violent storm where δ is very small, the rate of exchange of CO_2 must vary with weather conditions. Wind tunnel measurements have shown that CO_2 exchange is roughly proportional to the square of the wind velocity e.g. at 10 ms^{-1} wind speed, $\delta = 50 \mu$m and CO_2 exchange is twenty times greater than at 2 ms^{-1} where $\delta = 200 \mu$m.

The above discussion has assumed no chemical interaction of CO_2 with water. To obtain a full picture of CO_2 exchange the following reactions should be considered:

$$(CO_2)_g \rightleftharpoons (CO_2)_{aq} \tag{3.4}$$

$$(CO_2)_{aq} + H_2O \rightleftharpoons H_2CO_3 \tag{3.5}$$

$$H_2CO_3 \rightleftharpoons H^+ + HCO_3^- \quad K_1 = 4.47 \times 10^{-4} \, \text{mol} \, l^{-1} \tag{3.6}$$

$$HCO_3^- \rightleftharpoons H^+ + CO_3^{2-} \quad K_2 = 5.62 \times 10^{-11} \, \text{mol} \, l^{-1} \tag{3.7}$$

Almost all of the CO_2 in solution remains as $(CO_2)_{aq}$ with about 1 per cent as the unstable H_2CO_3. When this is taken into account the true value of K_1 is about 3×10^{-4} mol l^{-1}. In solutions of pH less than 4 the concentration of ionic species is negligible. Sea water, however, has a pH of about 8 so ionic species must be considered together with the rates of the equilibrium reactions that produce them. CO_2 exchange is influenced by the presence of ionic carbonate species as these have a higher mobility in solution than $(CO_2)_{aq}$. As the rate of CO_2 exchange is determined by the rate of transport of $(CO_2)_{aq}$, HCO_3^-, and CO_3^{2-} across the liquid laminar layer, the higher the concentration of ionic species the greater the rate of exchange. At a fixed pH and temperature the relative proportions of $(CO_2)_{aq}$, HCO_3^-, and CO_3^{2-} are fixed by equilibria (3.5), (3.6), and (3.7). Under these conditions the rate of CO_2 exchange depends upon the thickness δ of the liquid laminar layer. Fig. 3.4 shows clearly that the difference in mass of CO_2 exchanged per unit time for ionic and non-ionic species is small at low surface layer thickness but becomes significant as the surface layer thickness increases. This difference in rate of exchange may be expressed in terms of an exchange enhancement, α, where:

$$\alpha = \frac{\text{Exchange rate taking ionic equilibria into account}}{\text{Exchange rate assuming only } (CO_2)_{aq} \text{ participating}}. \tag{3.8}$$

With seawater, α is 1·62 for a surface layer thickness of 300 μm and 2·67 for a surface layer thickness of 600 μm. Under average conditions the open ocean has a surface layer thickness of less than 100 μm so this enhancement of CO_2 exchange is of little global consequence. This should be compared with the exchange of SO_2 into natural waters (see Chapter 5).

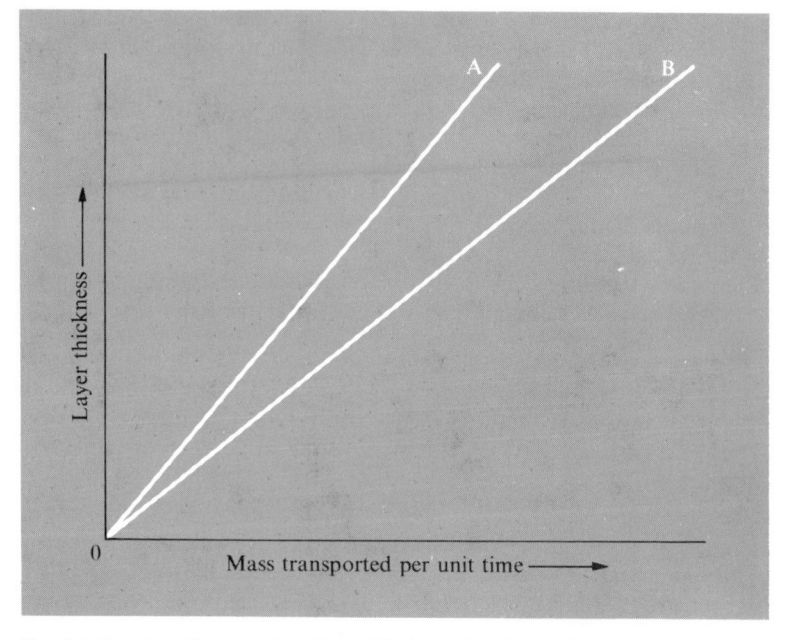

FIG. 3.4. Graph to illustrate the effect of liquid surface layer thickness on the mass of non-ionic (A) and ionic (B) CO_2 species transported in solution in unit time.

Carbon dioxide from combustion

Data which have been obtained for a number of sampling points remote from centres of man's activities indicate a constant increase in the atmospheric CO_2 concentration of about $1·0$ mg kg^{-1} y^{-1} (Fig. 3.5). A simple calculation of the annual mass of CO_2 emitted by combustion shows that the change in CO_2 concentration in the atmosphere due to combustion represents about one half of the emitted fossil fuel CO_2. The size of the oceanic reservoir and the solubility of CO_2 in water suggests that fossil fuel CO_2 should be removed more rapidly from the atmosphere. One of the reasons why this is not observed is that the rate of exchange of CO_2 into the large oceanic reservoir is relatively slow. Another is that the ocean is essentially a carbonate–bicarbonate buffer system and as a consequence a large increase in the atmospheric CO_2 partial pressure is necessary for a relatively small increase in the oceanic CO_2 concentration e.g. a 10 per cent increase in atmospheric CO_2 partial pressure results in a 0·6 per cent increase in oceanic CO_2 concentration.

Two major effects of increasing atmospheric CO_2 concentration are possible. Firstly, the increased CO_2 may affect photosynthesis; however, growth-chamber studies have shown that the present CO_2 concentrations

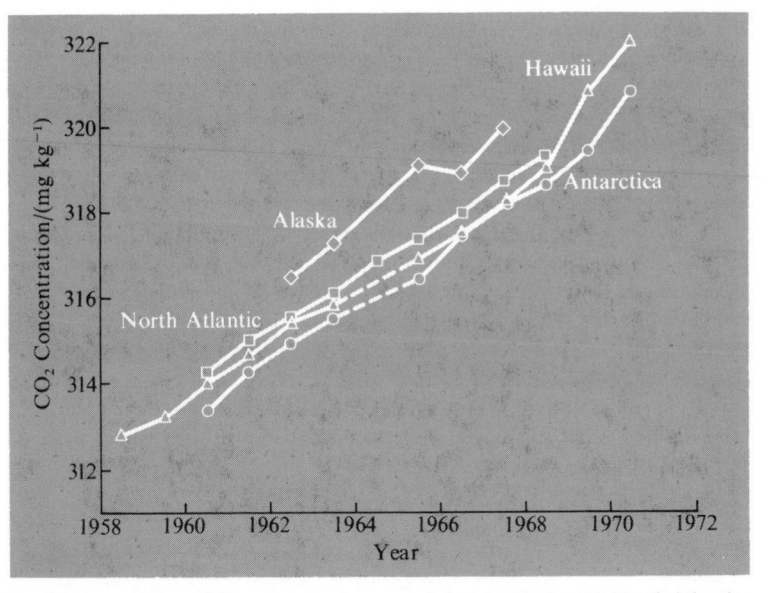

FIG. 3.5. Annual mean CO_2 concentration recorded at $\diamond$, Alaska; $\square$, North Atlantic; $\triangle$, Hawaii; $\bigcirc$, Antarctica, (after Garatt and Pearman 1972).

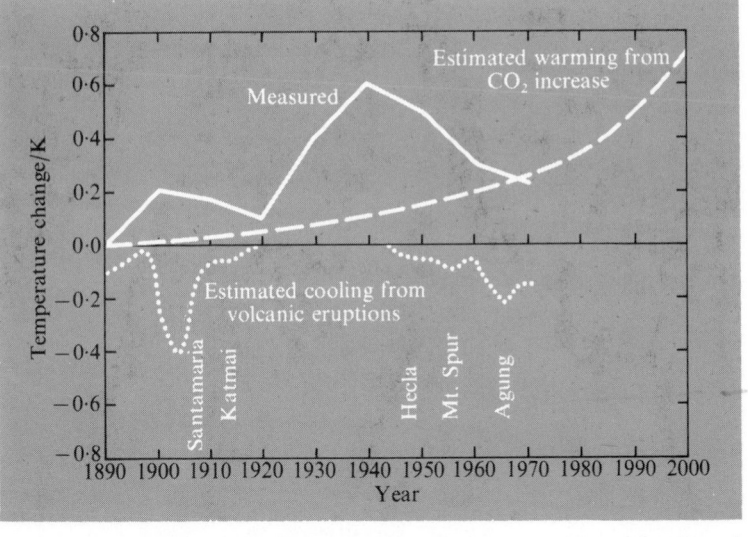

FIG. 3.6. Average Northern Hemisphere temperature changes together with estimated heating effects due to CO_2 increase and cooling due to volcanic dust (after Dyer 1972).

are below optimum for photosynthesis when other factors such as nutrient supply are optimum. Secondly, an effect on world climate may occur because of the greenhouse effect. Fig. 3.6 shows the calculated change in world temperature up to the year 2000 based on calculated emissions of CO_2 together with data on observed temperature changes during the 20th century. It can be seen that, to the present date, predicted temperature changes are smaller than natural changes so that the validity of the original predictions is in doubt. Some workers have suggested that the observed temperature increase from 1920 to 1940 was due to a CO_2 greenhouse effect which was, in 1940, decreased by increasing atmospheric turbidity from the solid particulate matter from combustion. The stratospheric aerosol from this source reflects radiant energy from the sun causing a decrease in atmospheric temperature. It is difficult to predict which of the two parameters (CO_2 greenhouse effect or atmospheric turbidity) will dominate in the future. Certainly, depending upon the parameters chosen it is possible to predict that in the coming years we will either 'freeze' or 'fry'.

4. Ozone

ALL of the important processes associated with ozone occur within the atmosphere, especially in the stratosphere. In earlier pages it has been shown that ozone is not uniformly distributed throughout the atmosphere but is confined largely to the stratosphere (Fig. 1.1). It has also been pointed out that the increase in temperature in the stratosphere is due to the absorption of solar ultraviolet radiation by ozone and that this absorption shields the surface of the earth from much of the solar ultraviolet thus reducing its damaging effects to terrestrial organisms. Reference to Fig. 3.1 will show that both oxygen and ozone absorb solar ultraviolet radiation. The photochemical reactions that are consequent upon this absorption form the basis of a photochemical equilibrium that maintains the ozone layer.

Ozone is formed in the following reactions:

$$O_2 + hv \rightarrow O + O \qquad (\lambda < 242\,nm) \qquad (4.1)$$

$$O_2 + O + M \rightarrow O_3 + M \qquad (4.2)$$

Reaction 4.2 is a three-body collision reaction where M is usually N_2 or O_2. The destruction of ozone is controlled by:

$$O_3 + hv \rightarrow O_2 + 0 \qquad (\lambda < 1180\,nm) \qquad (4.3)$$

$$O + O_3 \rightarrow 2O_2 \qquad (4.4)$$

There is thus an apparent equilibrium in the ozone region with the concentration of ozone remaining constant, however atmospheric mixing processes upset the equilibrium by removing some of the ozone to the troposphere where it is destroyed. The ozone concentration in the ozone region may reach $10\,mg\,kg^{-1}$ while at sea level in unpolluted areas it is as low as $0.01\,mg\,kg^{-1}$.

Almost all tropospheric ozone of natural origin is thought to come from the stratospheric ozone layer by exchange across the tropopause. In the troposphere ozone is chemically destroyed, primarily by contact with the earth's surface but also in clouds, and by gaseous and particulate trace substances. In the latter category are some of the reactions associated with photochemical smog which are dealt with in detail in Chapter 6.

Ozone of artificial origin is formed in the troposphere from photochemical reactions of pollutant gases that give rise to atomic oxygen (see Chapter 6). The atomic oxygen may then become involved in reaction 4.2 resulting in the formation of ozone. Concentrations of ozone up to about $0.2\,mg\,kg^{-1}$ can be obtained in localized areas from these reactions. In the past it has been thought that ozone from this source would only be found in regions of high sunlight intensity. Recently, however, evidence has become available that suggests that ozone levels as high as $0\cdot1$ mg kg^{-1} may be found in rural England in the summer months.

Recent observations of ozone concentrations in unpolluted areas, and experiments on artificial atmospheres suggest an alternative natural tropospheric source of ozone similar to the artificial source discussed in the preceding paragraph. The reactions producing this ozone may be described by the following general equation:

$$NO_2 + \text{olefin} + hv \longrightarrow O_3 + \text{other products.} \qquad (4.5)$$

Olefins occur naturally in the atmosphere, particularly as terpenes which cause the blue haze seen over forested areas, while nitrogen dioxide is produced largely by microbial action in the soil. It is most unlikely that sufficient ozone would be generated under these natural conditions to cause damage to our environment. On the other hand, pollutant ozone concentrations are sufficiently high to cause environmental damage.

Many plants are damaged by ozone concentrations as low as $0\cdot1$ mg kg^{-1}. The physiology of the damage is so well-known that plants can be used as indicators of ozone pollution by observing the nature of the lesions found on leaves of different plant species. The biochemical basis of the damage is unknown, although it is known that photosynthesis is inhibited by abnormal ozone concentrations. Some workers have attributed the inhibition of photosynthesis to a partial closing of the leaf pores (stomata) induced by ozone. The closure results in a lowering of CO_2 uptake and thus of the rate of photosynthesis.

Humans are also affected by low concentrations of ozone. The odour detection limit is $0\cdot02$ mg kg^{-1} to $0\cdot05$ mg kg^{-1}, irritation of the nose and throat occurs at $0\cdot05$ mg kg^{-1} and dryness of the upper respiratory mucosa is found at $0\cdot1$ mg kg^{-1}. Tests on rats have shown that the effect of ozone is not confined to the respiratory tract but biochemical processes are also affected e.g. it is known that ozone denatures proteins and thus enzymes. Relatively high concentrations of the gas would be needed for biochemical effects to occur to any extent, as the very reactive ozone has many surfaces on which it could be destroyed before it is absorbed into the body fluids.

The best-known example of damage to non-living materials by ozone is the deterioration of rubber in the presence of pollutant ozone and sunlight. In the Los Angeles County the deterioration of motor-vehicle tyres has been used to estimate ozone concentrations. Dyes that are susceptible to colour change under oxidizing conditions are affected by atmospheric ozone and it is suspected that asphalt is damaged by low ozone concentrations. The reader is referred to standard organic chemistry texts for the nature of the reaction between ozone and organic compounds.

Summary

Most atmospheric ozone is produced in the stratosphere and mixes slowly down to the troposphere where it is found at very low concentrations. Pollutant ozone arises almost exclusively from photochemical smog reactions (Chapter 6). Ozone concentrations from this source are often sufficient to damage vegetation, motor vehicle tyres, and asphalt and may, on occasion, cause irritation to the human respiratory tract. It is probably better to consider ozone effects on the enrivonment together with the effects of the other constituents of photochemical smog rather than in isolation.

5. Sulphur dioxide

SULPHUR is present in the atmosphere in at least three forms—SO_2, H_2S, and aerosol sulphate. On a global scale these species are involved in the sulphur cycle, a recent version of which is illustrated in Fig. 5.1. Two of the processes in this cycle may be taken as being atmospheric pollution processes. The production of SO_2 by the combustion of fossil fuels is an obvious pollution source. Less obvious is the emission of H_2S to the atmosphere by biological processes on land in coastal areas. Decomposition of organic waste products from man's activities accounts for some of this H_2S emission and illustrates clearly that water pollution (with which H_2S emission is usually associated) and air pollution are very closely related.

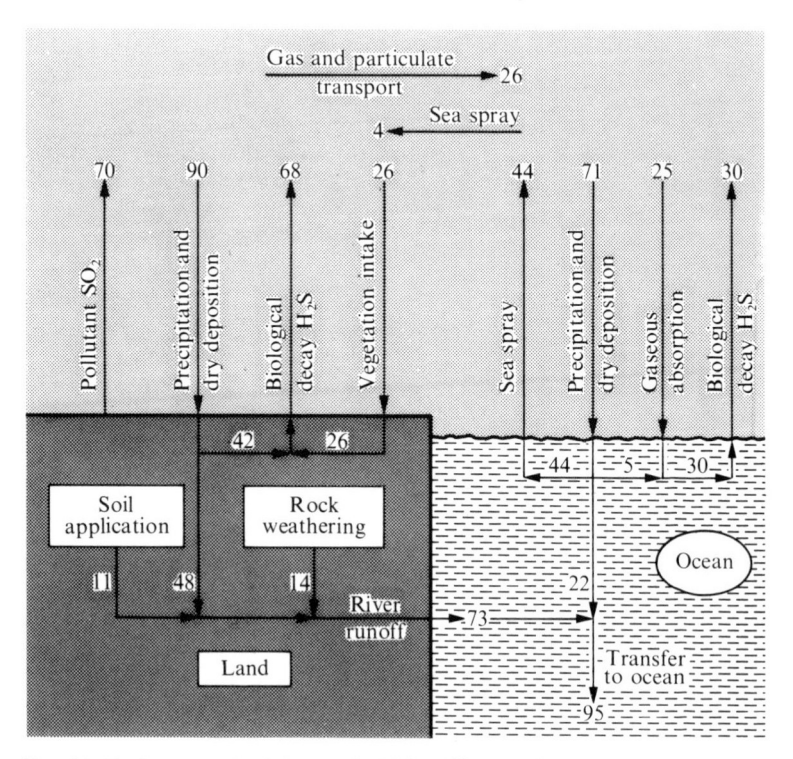

FIG. 5.1. Environmental sulphur cycle. Units 10^6 ton y^{-1} sulphur (after Robinson and Robbins 1970).

In gaseous form (SO_2 and H_2S) sulphur is quite rapidly transported in the atmosphere, but on oxidation transport is limited as sulphur is now present as solid sulphate salts or as sulphuric acid mist. Both the gaseous and the aerosol forms of atmospheric sulphur have detrimental effects on the environment. Vegetation is damaged, and some human respiratory complaints are worsened by quite low concentrations of atmospheric SO_2. Non-living materials in the environment tend to suffer damage from the acidic oxidation product of SO_2 rather than directly from the gas itself. Man especially expects his buildings and books to have a lifetime of centuries but in this time scale the amount of sulphuric acid formed on their surfaces is quite considerable. One thus finds building stone in the older industrial cities suffering from severe erosion and books stored in libraries of industrial cities deteriorating especially with respect to mechanical strength of the paper and bindings.

In this chapter the portions of the sulphur cycle that directly involve sulphur compounds emitted as pollutants to the atmosphere will be discussed, and the role of these compounds in the recent deterioration of the environment will be considered.

Sources

Table 5.1 lists the major sources of pollutant SO_2. The most prominent of these is the combustion of fossil fuels, all of which contain some sulphur compounds as 'contaminants' e.g. coal and fuel oil may have up to 3 per cent sulphur, while petrols usually contain about 0·05 per cent sulphur. Smelting of sulphide ores is the other major source.

TABLE 5.1

Annual pollutant SO_2 emissions for 1965 (after Robinson and Robbins (1970)

Source	Emission (10^6 tonnes)
Coal	102
Petroleum (combustion and refining)	28·5
Copper smelting	12·9
Lead smelting	1·5
Zinc smelting	1·3

It might be expected that such high total emissions of SO_2 would result in high atmospheric SO_2 concentrations. In localized areas near to large sources of pollution the SO_2 concentrations may reach as high as 1 mg kg^{-1} (about 3000 μg m^{-3}). The background SO_2 concentration has, however, been estimated to be in the range 0·3 to 1·0 μg m^{-3}.

The magnitude of the total H_2S source is unknown. All versions of the sulphur cycle deduce the H_2S source magnitude from the mass necessary to obtain a balanced cycle. No differentiation has been made between natural H_2S sources, and sources resulting from anaerobic metabolism of organic waste compounds from man and his activities. The average tropospheric concentrations of H_2S has been estimated to be $0.3 \, \mu g \, m^{-3}$ while concentrations of some Netherlands cities have reached more than $100 \, \mu g \, m^{-3}$ probably because of reducing conditions in the canals.

Oxidation of H_2S in the atmosphere

For some time now it has been accepted that H_2S is rapidly oxidized to SO_2 in the atmosphere. Evidence for such an atmospheric oxidation is scant. Possible modes of oxidation include homogeneous gas-phase oxidation by atomic and molecular oxygen and ozone, as well as heterogeneous oxidation in fog or cloud droplets by the same compounds. Reactions involving electronically-excited H_2S are unlikely as this molecule does not absorb solar radiation of the wavelengths reaching the troposphere.

Oxidation with oxygen at atmospheric temperatures and pressures proceeds immeasurably slowly. The reaction with ozone (5.1) can be measured although it too is very slow:

$$H_2S + O_3 \rightarrow SO_2 + H_2O \tag{5.1}$$

A more likely reaction is a chain reaction with atomic oxygen. The atomic oxygen required for this chain oxidation may be obtained from the photochemical dissociation of ozone (Chapter 4) or from photochemical smog reactions (Chapter 6).

No useful data are available on the solution oxidation of H_2S in fog or cloud droplets.

Oxidation of SO_2 in the atmosphere

One of the major steps in the sulphur cycle is the precipitation and dry deposition of sulphur from the atmosphere as sulphate in the rain, or as solid particulate matter. It is presumed that much of the atmospheric SO_2 is oxidized to sulphate and returned to the earth's surface in this form.

The rate at which SO_2 is oxidized determines its lifetime in the atmosphere provided that oxidation is the most important removal mechanism. A number of field experiments have been carried out to give the SO_2 oxidation rates summarized in Table 5.2.

The wide range of oxidation rates observed in the field imply variations in the conditions under which the measurements were made. A number of oxidation mechanisms are known and these could produce different rates under different conditions e.g. a photochemical mechanism would produce a high oxidation rate during the middle of the day while oxidation in fog droplets catalyzed by metal ions would be independent of sunlight. Most of

TABLE 5.2

Summary of SO_2 *oxidation rate studies in the atmosphere (after Urone and Schroeder 1969)*

Source of SO_2	Concentration of SO_2 on release from source $(mg\,kg^{-1})$	Relative humidity during measurement	Rate of SO_2 consumption in atmosphere $(\%\,ks^{-1})$
Ni smelter	0·1–1·0	—	0·6
Smelter	0·01–20·3	65–70%	190
Coal-fired power plant	2200	70–100%	1·7
Coal-fired power plant	2200	100%	8·3

the postulated mechanisms will now be discussed and their probable rates in the atmosphere compared.

Aqueous phase oxidation

A clear understanding of SO_2 oxidation in aqueous phases involves a knowledge of the solution chemistry of SO_2. The equilibria (5.2)–(5.6) outline the most important processes:

$$(SO_2)_g + H_2O \rightleftharpoons (SO_2)_{aq} \tag{5.2}$$

$$(SO_2)_{aq} + H_2O \rightleftharpoons H_2SO_3 \tag{5.3}$$

$$H_2SO_3 + H_2O \overset{K_1}{\rightleftharpoons} H_3O^+ + HSO_3^-; \quad K_1 = 1\cdot6 \times 10^{-2} \tag{5.4}$$

$$HSO_3^- + H_2O \overset{K_2}{\rightleftharpoons} H_3O^+ + SO_3^{2-}; \quad K_2 = 1\cdot0 \times 10^{-7} \tag{5.5}$$

$$2HSO_3^- \overset{K_3}{\rightleftharpoons} S_2O_5^{2-} + H_2O; \quad K_3 = 7 \times 10^{-2}\,mol^{-1}. \tag{5.6}$$

A solution of SO_2 in water thus contains hydrated SO_2, H_2SO_3, HSO_3^-, SO_3^{2-}, and $S_2O_5^{2-}$ in proportions that vary with pH and concentration. Sulphurous acid, H_2SO_3, is unknown as a free acid and because of this it is often represented as $H_2O \cdot SO_2$. Rainwater usually has a pH of 4 to 6 thus the predominant ion in atmospheric water is HSO_3^-.

Oxidation of a solution of sulphur compounds in oxidation state $+4$ as represented by eqns (5.2) to (5.6) may be direct to oxidation state $+6$ (sulphate) or it may pass through oxidation state $+5$ (dithionate):

$$(S^{IV})_{aq} \xrightarrow{O_2} (S^{VI})_{aq}$$
$$\underset{O_2}{\searrow} (S^{V})_{aq} \underset{O_2}{\nearrow}$$

Sulphate is the only higher oxidation state sulphur compound found in atmospheric water. Dithionate has never been reported but this may be because no great effort has been made to detect it.

The direct oxidation of $(S^{IV})_{aq}$ by atmospheric oxygen proceeds very slowly, if at all. The presence of metal ion catalysts, especially Mn^{2+}, Fe^{3+}, and Cu^{2+} results in a rapid oxidation. The general oxidation pathway has recently been deduced from flash photolysis studies:

Initiation:

$$SO_3^{2-} \xrightarrow{hv} SO_3^- + e_{aq}^- \tag{5.7}$$

$$SO_3^- + O_2 \rightarrow SO_5^- \tag{5.8}$$

Propagation:

$$SO_5^- + SO_3^{2-} \rightarrow SO_4^- + SO_4^{2-} \tag{5.9}$$

$$SO_4^- + SO_3^{2-} \rightarrow SO_4^{2-} + SO_3^- \tag{5.10}$$

Termination:

$$SO_5^- + SO_5^- \rightarrow \text{products} \tag{5.11}$$

$$SO_4^- + SO_4^- \rightarrow \text{products} \tag{5.12}$$

In solutions of lower pH, where HSO_3^- predominates, reaction (5.10) becomes:

$$SO_4^- + HSO_3^- \rightarrow HSO_4^- + SO_3^- \tag{5.13}$$

Reaction (5.13) is about 2·5 times faster than reaction (5.10) and thus explains an increased oxidation rate at a pH approximating to the value of pK_1 for reaction (5.4).

Reaction (5.7) has been given for flash photolytic initiation. In thermal autoxidation this reaction is replaced by some other electron transfer e.g.

$$SO_3^{2-} + Cu^{2+} \rightarrow SO_3^- + Cu^+ \tag{5.14}$$

It is difficult, however, to provide a reaction similar to (5.14) for the case where Mn^{2+} acts as a catalyst, as Mn^I exists in only a few complexes in non-aqueous solvents. A mechanism involving Mn^{II}–SO_2 complexes may be used:

$$SO_2 + Mn^{2+} \rightarrow [MnSO_2]^{2+} \tag{5.15}$$

$$2[Mn\cdot SO_2]^{2+} + O_2 \rightarrow [(Mn\cdot SO_2^{2+})_2O_2] \rightarrow 2[MnSO_3]^{2+} \tag{5.16}$$

$$[MnSO_3]^{2+} + H_2O \rightarrow Mn^{2+} + HSO_4^- + H^+ \tag{5.17}$$

Reaction (5.16) is probably rate-determining as Mn-bound oxygen is rearranged to S-bound oxygen in the oxidation step.

Similar mechanisms have been postulated for other metal ion catalysts. Whether the radical chain mechanism or the metal complex mechanism or a combination of both is correct awaits further experimental elucidation.

An important observation from the point of view of atmospheric chemistry is that the rate of oxidation of SO_2 decreases as the pH falls below the value of pK_1 for reaction (5.4). As the oxidation of SO_2 results in the formation of H_2SO_4, which is a stronger acid than H_2SO_3, the pH of the reaction solution decreases as oxidation proceeds. The rate of oxidation thus decreases with time. In order to provide a mechanism to explain the relatively high rate of SO_2 oxidation in atmospheric water some workers have suggested that the buffer capacity of dissolved CO_2 and NH_3 may be sufficient to maintain solution pH at a level where oxidation is quite rapid. The effect of ammonia in solution was found to be quite dramatic—with an SO_2 concentration of $20 \mu g m^{-3}$ and no NH_3 a sulphate concentration of 0·5 $\mu g m^{-3}$ was calculated for 24 hours oxidation, while in the presence of 5 $\mu g m^{-3}$ NH_3 the sulphate concentration was calculated as about 60 $\mu g m^{-3}$ for the same time. The concentration parameters for both SO_2 and NH_3 are reasonable for atmospheres found in populated areas.

Ammonium sulphate, the product of SO_2 oxidation in the presence of NH_3, has been identified as a major constituent of the atmospheric aerosol. Because of its deliquescent behaviour it is of considerable importance in the formation of hazes and mists which have a marked influence on visual range.

Photo-oxidation

Here we are to consider reactions of SO_2 that, in the gas phase and in the presence of solar radiation, produce SO_3 and hence H_2SO_4. The photochemical dissociation of SO_2 (reaction 5.18) is not possible under atmospheric conditions as it has an energy requirement of 565 kJ mol^{-1} which cannot be satisfied by the absorption of solar radiation.

$$SO_2 + h\nu \rightarrow SO + O \qquad (5.18)$$

An alternative mode of photo-oxidation of SO_2 is through an electronically-excited state of SO_2 produced when SO_2 absorbs solar radiation. The absorption spectrum of SO_2 shows two bands in the wavelengths of tropospheric solar radiation. A weak absorption with a maximum at 388 nm produces a triplet state ($^3SO_2$) while a strong absorption with a maximum at 294 nm produces a singlet state ($^1SO_2$). The reactions of these electronically-excited states of SO_2 in pure SO_2 are summarized in reactions (5.19)–(5.27).

$$SO_2 + h\nu \rightarrow {}^1SO_2 \qquad (294 \text{ nm}) \qquad (5.19)$$

$${}^1SO_2 + SO_2 \rightarrow (2SO_2) \qquad (5.20)$$

$${}^1SO_2 + SO_2 \rightarrow {}^3SO_2 + SO_2 \qquad (5.21)$$

$${}^1SO_2 \rightarrow SO_2 + h\nu_f \qquad (5.22)$$

$${}^1SO_2 \rightarrow SO_2 \qquad (5.23)$$

$${}^1SO_2 \rightarrow {}^3SO_2 \qquad (5.24)$$

$${}^3SO_2 \rightarrow SO_2 + h\nu_p \qquad (5.25)$$

$$^3SO_2 \rightarrow SO_2 \qquad (5.26)$$

$$^3SO_2 + SO_2 \rightarrow (2SO_2) \qquad (5.27)$$

Reactions (5.21) and (5.24) are intercrossing reactions by which $^3SO_2$ may be formed from $^1SO_2$. They are very important as it has been shown that $^3SO_2$ is the reactive species in SO_2 photo-oxidation and, as SO_2 has only a weak absorption band for the direct formation of $^3SO_2$, they provide a pathway for the formation of significant amounts of $^3SO_2$.

Reactions (5.22), (5.23), (5.25), and (5.26) represent the decay to ground-state SO_2 from both excited states either by the emission of radiation or by radiationless decay. The product, $(2SO_2)$, of reactions (5.20) and (5.27) can be any chemical product that does not regenerate an excited state of SO_2 and may be SO_3 or a compound that produces SO_3 e.g. the very reactive tetroxide, SO_4.

In the atmosphere SO_2 is only a trace constituent hence other interactions similar to (5.20), (5.21), and (5.27) must also be considered:

$$^1SO_2 + M \rightarrow SO_2 + M \qquad (5.28)$$

$$^1SO_2 + M \rightarrow {}^3SO_2 + M \qquad (5.29)$$

$$^3SO_2 + M \rightarrow products \qquad (5.30)$$

$$^3SO_2 + M \rightarrow SO_2 + M \qquad (5.31)$$

where M is some molecule other than SO_2, usually N_2, O_2, or H_2O because of their high abundance, or olefins because of their very high reactivity. Reaction (5.30) is the most important reaction for the production of SO_3. The exact nature of this reaction is not clear, however the following postulated reactions have considerable support from workers in the field:

$$^3SO_2 + O_2 \rightarrow SO_3 + O \qquad (5.32)$$

$$O + O_2 \rightarrow O_3 \qquad (5.33)$$

Ozone has never been found in SO_2 photo-oxidation, perhaps because of the analytical difficulty of its determination in the presence of SO_2.

Oxidation on aerosols

Although SO_2 has been shown to be rapidly sorbed by aerosols of Fe_3O_4, Al_2O_3, and CaO the subsequent reactions of the gas are not known in any detail. Sulphate has been found in the solid particulate material in the atmosphere but it has not been shown that it was formed *in situ* on the aerosol by oxidation of SO_2. It could well have arisen from coagulation of a H_2SO_4 aerosol droplet with a solid particulate aerosol.

Comparison of the oxidation mechanisms

When making a comparison of the mechanisms for SO_2 oxidation in the atmosphere a wider range of parameters must be considered than were considered in laboratory studies. For instance, the photochemical mechanism will obviously have a zero oxidation rate at night and the rate will also vary with the intensity of solar radiation during the day. Taking these factors into consideration an SO_2 oxidation rate of about 0·15 per cent ks^{-1} integrated over a twenty-four hour period can be deduced for photo-oxidation.

In the case of SO_2 oxidation in droplets in the atmosphere consideration must be given to the rate of dissolution of SO_2 into the droplets, the rate of dissolution of metal oxides to produce metal ions for catalysis, and to the relative humidity of the atmosphere which controls droplet formation. No estimates have been made taking into account the rate of production of metal ions for catalysis. All calculations have been based on the assumption that all of the metal under consideration is available in soluble form to assist SO_2 oxidation. The maximum oxidation rates calculated on this basis occur under fog conditions and may reach a rate of 0·6 per cent ks^{-1} with Mn^{II} as a catalyst. It is possible to calculate rates of up to 25 per cent ks^{-1} with Fe^{III} as a catalyst in a chimney plume. The oxidation rates reported in Table 5.1 are thus adequately explained by this mechanism.

The mean residence time of SO_2 in the atmosphere has been derived from data collected at monitoring stations in various parts of the world. The results vary from twelve hours to six days. It can thus be seen that photochemical oxidation and solution oxidation separately, or together, could account for the longer observed residence times. The shorter residence times must be influenced by factors other than SO_2 oxidation e.g. absorption of SO_2 into natural waters, deposition of SO_2 on to vegetation and soils, etc.

Deposition of SO_2 into natural waters

In Chapter 3 the basic physical concepts for gaseous exchange between the gas phase and the aqueous phase were discussed with reference to CO_2. In that discussion the influence of the gas-phase boundary layer was considered to be insignificant. Experiments with SO_2 have shown that the exchange coefficient is constant above pH 6 while it decreases markedly with pH below pH 6. This implies that the exchange coefficient is influenced by the change of ionic species in solution and it can be shown that, at most pH values, diffusion through the gas-phase boundary layer must be considered.

The resistance R to SO_2 exchange from the gas phase to solution is, by analogy with Ohm's Law, the sum of the resistance to exchange through the gas-phase boundary layer r_g and the resistance to exchange through the solution phase boundary layer r_1, i.e.

$$R = r_g + r_1. \tag{5.34}$$

Continuing the analogy with electricity we find that the exchange coefficient is analogous with conductance hence eqn (5.34) may be written:

$$\frac{1}{k} = \frac{1}{k_l} + \frac{1}{Hk_g} \qquad (5.35)$$

where k is the overall exchange coefficient, k_l is the exchange coefficient for transport through the solution phase, and k_g is the exchange coefficient for transport through the gas phase. H, Henry's Law constant, is applied to bring the exchange coefficients to the same concentration basis.

The value for k_g depends solely upon windspeed and may thus vary from 0.3 cm s^{-1} in calm conditions up to 1.5 cm s^{-1} in turbulent conditions. On the other hand, k_l is greatly influenced by the ionic species in solution which vary markedly with pH. At pHs where ionic species predominate the enhancement α, (see Chapter 3), to the exchange coefficient for aquated molecular SO_2 alone is very large indeed, e.g. at pH 2 where ionic S^{IV} species are in low concentration α is about 3 whereas at pH 6 α is about 3000. The overall effect of this can be seen in Table 5.3. At pH 2.8, under the wind

TABLE 5.3

Calculated resistance to SO_2 exchange into aqueous solutions of varying pH
(after Liss 1971)

Solution pH	Ionic enhancement factor, α	Liquid-layer resistance, r_l (cm s^{-1})	Gas-layer resistance, r_g (cm s^{-1})	Total resistance, R (cm s^{-1})
2	2.7	1.00	0.25	1.25
3	18	0.15	0.25	0.40
4	169	0.02	0.25	0.27
5	1376	0.002	0.25	0.25
6	2884	0.001	0.25	0.25
7	2966	0.001	0.25	0.25
8	2967	0.001	0.25	0.25
9	2967	0.001	0.25	0.25

conditions chosen for the calculation, $r_l = r_g$. Below pH 2.8 r_l is dominant, and above r_g is dominant. Most natural waters are of pH 4–9 hence it can be concluded that at almost all environmental air–water interfaces, SO_2 exchange rate is controlled by diffusion through the gas-phase boundary layer. Water vapour exhibits similar behaviour, while O_2 and CO_2 exchange with natural waters are controlled by diffusion through the liquid phase boundary layer.

As SO_2 has a high solubility (at 15°C 45 vols SO_2 dissolve in 1 vol H_2O) virtually all SO_2 reaching the surface of natural waters will dissolve. Sea water has a pH of about 8 and is buffered by the carbonate–bicarbonate system. It can thus be concluded that the oceans are an important sink for SO_2. Recently a direct measurement of the rate of absorption of SO_2 by seawater has been made. Under the most turbulent conditions the value for the total resistance to exchange was 0.7 s cm^{-1}. This value may be used in the following manner to estimate the total uptake of SO_2 from the atmosphere by the oceans.

The total resistance R is related to the rate of absorption by:

$$R = \frac{1}{V_g} \tag{5.36}$$

where V_g, the velocity of deposition, is defined as

$$V_g = \frac{M}{[SO_2]_g \times t} \tag{5.37}$$

where M = Mass of SO_2 (μg) sorbed per unit area (cm^2)
    $[SO_2]_g$ = Gas-phase concentration of SO_2 ($\mu\text{g cm}^{-3}$)
      t = Time of gas exposure to solution (s).

Data necessary for calculation:
  Number of seconds in one year = 3.156×10^7 s
  Average concentration of S as SO_2 over oceans = $2.5 \times 10^{-7} \, \mu\text{g cm}^{-3}$

$$V_g = \frac{1}{R} = 1.4 \text{ cm s}^{-1}$$

Area of oceans = $3.7 \times 10^{18} \text{ cm}^2$
Mass of S sorbed as SO_2 = $1.4 \times 2.5 \times 10^{-7} \times 3.156 \times 10^7 \times 3.7 \times 10^{18} \, \mu\text{g}$
                   = 41×10^6 tonnes.

This represents a greater absorption of SO_2 by the oceans than is given for the sulphur cycle (Fig. 5.1). The difference arises only from the value of V_g used. The sulphur cycle figure was obtained using an estimated V_g of 0.9 cm s^{-1} while the above calculation used a measured V_g of 1.4 cm s^{-1}. Greater SO_2 absorption by the ocean suggests a lower emission of H_2S or sea-spray sulphate necessary to balance the cycle. The former is more attractive as it has often been considered that H_2S would be rapidly oxidized in sea water and thus be unavailable for exchange to the atmosphere.

Deposition of SO_2 to vegetation and soil

Although most plants are damaged by SO_2 when exposed to concentrations of about 1 mg kg^{-1} for several hours, almost all of the world's vegetation receives non-damaging concentrations of SO_2. Plant uptake of SO_2 is almost completely through the leaf pores (stomata) which control gas exchange between the interior of the leaf and the atmosphere. The stomata are under the physiological control of the plant and open or close depending on the plant's need for CO_2 or water exchange. When the atmosphere is at a low relative humidity the stomata close to conserve water while high relative humidity and high sunlight intensity cause stomatal opening so that CO_2 exchange and photosynthesis are promoted. Maximum SO_2 uptake is found at 100 per cent relative humidity in the atmosphere and corresponds to maximum stomatal opening.

As was the case with SO_2–sea water exchange it is possible to express SO_2 uptake by leaves as a resistance to total gaseous exchange, R.

$$R = r_g + r_s + r_{mes} \tag{5.38}$$

R and r_g are as defined in (5.34), r_s is the resistance to exchange through the stomata, and r_{mes} is the resistance to exchange at the damp cell (mesophyll) surfaces within the leaves.

The minimum value for R has been found to be 2.8 s cm^{-1}. Values for r_g and r_s can be calculated for a given set of experimental conditions. The sum of these two resistances under conditions of maximum SO_2 exchange approaches the value of R. This implies that the resistance to SO_2 exchange at the damp mesophyll cell surfaces is very low and, by analogy with r_1 (eqn 5.34), is not controlled by chemical reaction within the mesophyll cells. The reverse is true for CO_2, where it is found that r_{mes} is an important contributor to R.

The reciprocal of R gives the velocity of deposition (V_g) for maximum SO_2 exchange to plant leaves. This has a value of 0.36 cm s^{-1}. In order to calculate the total deposition of SO_2 to plant leaves per unit area of land it is necessary to estimate the total leaf area per unit area of land. This varies with plant species but a factor of four times the leaf area (top plus underside) per unit area of land appears to be the average. The overall value of V_g for SO_2 deposition to plant leaves, per unit area of land, is thus about 1.5 cm s^{-1}. A calculation similar to that carried out for SO_2 absorption by sea water leads to a value for annual SO_2 intake by vegetation similar to that given in the Sulphur Cycle (Fig. 5.1).

It is also possible to calculate that vegetation is able to be supplied annually with 50 mg m^{-2} of S for each $1 \text{ } \mu\text{g m}^{-3}$ of SO_2 in the atmosphere. This would be marginally sufficient as the sole source of sulphur for a wide range of crops. Many soils in the world are deficient in sulphur hence the transfer of sulphur in fuels directly to plants in the form of atmospheric SO_2 could be of some importance to world agriculture especially in regions where the application of chemical fertilizers is not carried out for economic reasons.

The direct absorption of SO_2 by soils is not included in the Sulphur Cycle. It is known that soils absorb SO_2 and that the rate depends very much upon the moisture content of the soil. At high moisture contents the SO_2 absorption rate is maximal. Micro-organisms also have an effect, causing an increase in absorption rate as their population increases. In calculating the mass of SO_2 absorbed by vegetation it was assumed that the whole of the surface of the earth was covered by plants. The uptake of SO_2 by soils is probably accounted for in the total sulphur cycle by this simplification.

Damaging effects of SO_2

Effects on man

Few data are available on the effects of SO_2 air pollution on human health. The classic epidemiological study of the London smog of 5–8 December 1952 showed an excess of 3500–4000 deaths above the predicted value. In this smog high concentrations of particulate matter as well as SO_2 were measured. It is generally accepted that the combined effect of both constituents on the respiratory tract was responsible for the excess deaths.

Laboratory data indicate that SO_2 has the potential of slowing down ciliary movements in the respiratory tract. The cilia act to clear micro-organisms and toxic particles from the respiratory tract. If these irritants reach the lungs they may cause acute respiratory problems. There is no evidence that SO_2 at air pollutant concentrations in the absence of high concentrations of particulate material can cause adverse effects to humans in a normal state of health.

Effects on plants

It is often found with gaseous air pollutants that plants are damaged at much lower concentrations than those at which human health is affected. Sensitivity to SO_2 varies with plant species e.g. alfalfa, barley, cotton and wheat are listed as sensitive while potato, onion, corn and maple are listed as resistant. It appears that even the most sensitive species shows no visible response to an exposure of less than $100\ \mu g\ m^{-3}$ of SO_2, for an indefinite period. Continuous chronic exposure to SO_2 concentrations of as low as $300\ \mu g\ m^{-3}$ lead to leaf damage in some sensitive species, while short-time acute exposures may cause damage to leaves at as low as $700\ \mu g\ m^{-3}$ SO_2.

Sufficient knowledge has now accumulated to enable the identification of SO_2 damage by the physiological state of the leaves e.g. discolourations, necroses etc. Little is known of the biochemical basis for the damage although recently an aldehyde–hydrogensulphite adduct has been isolated from rice plants treated with SO_2.

$$\underset{\overset{|}{\text{HC}=\text{O}}}{\text{COO}^-}\ \text{glyoxalate}\ \xrightarrow{\text{HSO}_3^-}\ \underset{\overset{|}{\underset{\overset{|}{\text{OH}}}{\text{HC·SO}_3^-}}}{\text{COO}^-}\ \begin{array}{l}\text{glyoxalate}\\ \text{hydrogensulphite}\end{array} \quad (5.39)$$

Glyoxalate hydrogensulphite is a competitive inhibitor of the enzyme glycolic oxidase which catalyzes the oxidation of glycolate:

$$\underset{\overset{|}{\underset{\overset{|}{\text{glycolate}}}{\text{CH}_2\text{OH}}}}{\text{COO}^-}\ +\ \text{O}_2\ \xrightarrow{\text{glycolic oxidase}}\ \underset{\overset{|}{\underset{\overset{|}{\text{glyoxalate}}}{\text{CHO}}}}{\text{COO}^-}\ +\ \text{H}_2\text{O}_2 \quad (5.40)$$

Reaction (5.40) is part of the glycolic pathway which is associated with the process of photorespiration in plants. Independent evidence indicates that glycolic oxidase inhibition occurs in barley plants treated with SO_2. The importance of these observations awaits further experimental elucidation.

Effects on building materials

Limestone and marble have been used as building materials for many centuries. The damage to limestone in particular is obvious in industrial cities where severe erosion of limestone building blocks and ornamentation may be seen.

The mass of SO_2 absorbed by limestone is increased as relative humidity increases. The absorbed gas is oxidized to sulphate and becomes part of the $CaCO_3$ matrix. The molecular volume of $CaSO_4$ is greater than that of $CaCO_3$ hence mechanical stresses on the molecular scale arise. The accumulated effect of these stresses is to cause flaking-off of the limestone. Further, $CaSO_4$ has a higher solubility in rain water than $CaCO_3$ ($209\,mg\,ml^{-1}$ and $1.4\,mg\,ml^{-1}$ respectively) and is thus readily leached out.

Works of art, especially frescoes, are susceptible to SO_2 damage by the same mechanism. The true frescoe is a pigmented lime plaster which may also be converted to $CaSO_4$ in the presence of SO_2. Marble is less susceptible to SO_2 attack probably because of its low porosity.

Unpainted timber absorbs SO_2 and this may cause some degradation due to the reaction of sulphite with lignins cf. the sulphite-pulping process for the production of cellulose from wood. Damage by other weathering processes probably outweighs that due to SO_2. Painted timbers may also be attacked by SO_2 as it has been found that many paint films are permeable to SO_2.

Effects on paper

Of the many objects that are stored by man probably the most important are books. The yellowing and loss of mechanical strength of paper is accelerated in books stored in industrial cities. SO_2 has been clearly shown to be the major contributor to the degradation of paper stored in these places.

The usual phenomenon of increased SO_2 sorption with increased atmospheric humidity is also found with paper. Studies with radioactive SO_2 have shown that high concentrations of SO_2 are found about metallic impurities left in the paper from the manufacturing process. It is supposed that SO_2 is rapidly oxidized to H_2SO_4 at these points and that the basic degradation process is one of acid hydrolysis of cellulose. In addition, it is thought that the formation of lignosulphonic acids by the reaction between SO_2 in the surface moisture on the paper and lignins within it, also has an effect.

The handling of paper results in sweat deposits being left on the paper. Such deposits show a great capacity to sorb SO_2 hence damage to the outer edges of books may not only be due to abrasive actions but also to high

H_2SO_4 concentrations within sweat deposits. Similar effects occur on wall-papers but wallpapers are usually changed for social reasons before significant damage can be seen.

Effects on leather

Damage to leather by SO_2 has been known since at least 1843 when Faraday suggested that the deterioration of the leather armchairs of his London club was due to SO_2. Leather bookbindings also suffer SO_2 damage which takes the form of cracking and loss of flexibility and mechanical strength.

It has been shown that SO_2 uptake by leather is almost completely controlled by the rate of SO_2 diffusion to the surface. At the surface oxidation results in the formation of H_2SO_4 which may cause acid hydrolysis of leather protein. The amount of SO_2 sorbed by leather can be greatly reduced by providing a surface coating of nitrocellulose or polyurethane.

Effect on metals

Observations on the rate of corrosion of test panels of many metals exposed to the atmosphere at a variety of sites have shown that, with most metals, greatest corrosion occurs in industrial atmospheres. In these conditions the rate of corrosion correlates very closely with the SO_2 concentration in the atmosphere and with the time of wetness of the metal surfaces.

The atmospheric corrosion of iron and steel may be explained by an electrochemical mechanism. In the atmosphere iron and steel are always covered with a thin layer of Fe_3O_4 which itself is covered by a film of its oxidation product FeOOH (Fig. 5.2). At the surface of the pure metal the anodic reaction (5.41) occurs, while at the Fe_3O_4/FeOOH interface cathodic reduction followed by atmospheric oxidation occurs (reactions 5.42 and 5.43).

$$Fe \rightarrow Fe^{2+} + 2e \qquad (5.41)$$

$$Fe^{2+} + 8FeOOH + 2e \rightarrow 3Fe_3O_4 + 4H_2O \qquad (5.42)$$

$$3Fe_3O_4 + 0.75O_2 + 4.5H_2O \rightarrow 9FeOOH \qquad (5.43)$$

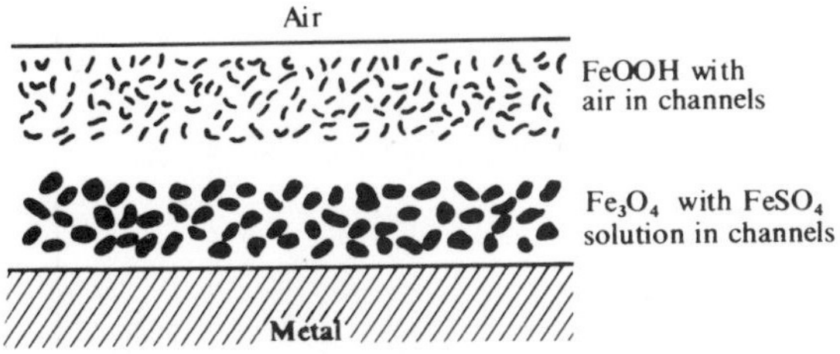

FIG. 5.2. Schematic representation of the electrochemical mechanism of atmospheric rusting of iron in the presence of SO_2 (after Evans 1972).

The overall effect is to increase the amount of rust, FeOOH, by one eighth by transferring iron from the metal to the surface rust layer.

As the anodic and cathodic sites for the corrosion process are separate in space, conductors are needed to complete the electrochemical circuit. It is supposed that magnetite, Fe_3O_4, is the electronic conductor and that $FeSO_4$ which exists in solution in the Fe_3O_4 layer acts as the ionic conductor. The role of SO_2 in the corrosion process is thus to provide SO_4^{2-} by its oxidation after sorption on the metal. The time of wetness relationship is explained by the necessity for $FeSO_4$ to remain in solution in the Fe_3O_4 layer so that the electrochemical circuit is completed. When the layer dries out the circuit is disrupted and corrosion ceases.

Aluminium is susceptible to attack in both industrial and marine atmospheres. The attack in industrial atmospheres is thought to be due to the formation of H_2SO_4 from SO_2 absorbed on the surface. The acid breaks down the oxide film that normally provides a protection against atmospheric attack, and corrosion proceeds. A similar process occurs on the surface of zinc where the protective film of basic zinc carbonate is dissolved by H_2SO_4 formed from SO_2 absorbed on the zinc surface.

6. Nitrogen oxides and photochemical smog

NITROGEN oxides play a very important role in the formation of the type of smog that was first recognized in the Los Angeles district, and which is now referred to as photochemical smog. This type of smog is typified by photochemical reactions brought about by the action of solar radiation on motor vehicle exhausts. The most damaging constituents of the smog are nitric oxide, nitrogen dioxide, ozone, and peroxoacyl nitrates. Vegetation is damaged by very low concentrations of some of these constituents while slightly higher concentrations cause unpleasant effects to humans, especially to the respiratory system.

Nitrogen forms oxides corresponding to each of its known oxidation states (Table 6.1). Of these oxides, only N_2O, NO, and NO_2 appear at measurable concentrations in the unpolluted atmosphere. The equilibria involving N_2O_3, N_2O_4, and N_2O_5 are all heavily in favour of dissociation at atmospheric temperatures and partial pressures.

TABLE 6.1

Oxides of nitrogen

Oxide	Formula	Stability in the atmosphere
Dinitrogen oxide	N_2O	Stable gas
Nitrogen oxide	NO	Stable gas
Dinitrogen trioxide	N_2O_3	Unstable gas $N_2O_3 \rightleftharpoons NO + NO_2$
Nitrogen dioxide	NO_2	Stable gas
Dinitrogen tetroxide	N_2O_4	Unstable gas $N_2O_4 \rightleftharpoons 2NO_2$
Dinitrogen pentoxide	N_2O_5	Unstable gas $N_2O_5 \rightleftharpoons N_2O_3 + O_2$
Nitrogen trioxide	NO_3	Unstable gas (never isolated)

Dinitrogen oxide (nitrous oxide)

This gas was first identified in the atmosphere by its infrared absorption spectrum as recently as 1939. The mean atmospheric concentration deduced from the infrared spectral data is $0.25 \, mg \, kg^{-1}$. The concentration of N_2O remains reasonably constant up to the tropopause and then decreases with altitude because of photodissociation reactions (6.1) and (6.2).

$$N_2O + h\nu \rightarrow N_2 + O \qquad (\lambda < 337 \, nm) \qquad (6.1)$$

$$N_2O + h\nu \rightarrow NO + N \qquad (\lambda < 250 \, nm) \qquad (6.2)$$

Photodissociation is not found to any extent in the troposphere as the solar radiation spectrum in that layer has a sharp cut-off in the ultraviolet region at 390 nm due to absorption in the ozone layer (see Fig. 3.1).

The principal source of atmospheric N_2O is the soil. Micro-organisms in the soil are able to degrade protein nitrogen to nitrogen gas and N_2O. The maximum rate of production of N_2O thus depends upon optimum conditions for the activity of soil micro-organisms, and a plentiful supply of protein. These conditions vary from season to season, hence it is not surprising that a seasonal variation in the concentration of N_2O in the atmosphere is found.

Micro-organisms are also capable of reducing N_2O under anaerobic conditions. It is thought that N_2O that diffuses to the deep ocean is destroyed in this manner. The most important processes for the destruction are, however, the stratospheric reactions (6.1) and (6.2). The normal cycle of atmospheric N_2O is thus production in the soil followed by diffusion to the stratosphere where photodissociation occurs. The calculated mean residence time for N_2O is subject to some conjecture. Values between 4 years and 70 years have been obtained.

N_2O is not recognized as an air pollutant. Its importance in the field of air pollution chemistry is in its photodissociation to NO which is an important pollutant gas.

Nitrogen oxide (nitric oxide) and nitrogen dioxide

These two oxides will be discussed together because of their relationship through equilibrium (6.3)

$$2NO + O_2 \underset{k_{-1}}{\overset{k_1}{\rightleftharpoons}} 2NO_2 \rightleftharpoons N_2O_4 \tag{6.3}$$

In the solid state nitrogen dioxide exists entirely as colourless N_2O_4 while in the vapour state at $100°C$ the composition at equilibrium is 90 per cent NO_2 and 10 per cent N_2O_4. For the purposes of further discussion nitrogen dioxide will be regarded as being the brown gaseous NO_2.

The production of NO_2 from NO in the atmosphere takes place rather slowly as reaction (6.3) is second order in NO concentration. At an NO concentration of $0.1\ mg\ kg^{-1}$ the half life for the reaction is about 4 Ms, which is in marked contrast to the almost immediate production of brown NO_2 fumes from high concentrations of NO. The NO_2 produced absorbs strongly in the ultraviolet region, dissociating to NO and atomic oxygen (reaction 6.4). Reaction (6.5) accounts for most of the atomic oxygen produced. The overall effect is thus to produce equal concentrations of NO and O_3 from the NO_2 present in the atmosphere. These compounds react together (reaction 6.6) thus completing a cyclic reaction,

$$NO_2 + h\nu \rightarrow NO + O \quad (\lambda < 380\ nm) \tag{6.4}$$

$$O + O_2 + M \rightarrow O_3 + M \tag{6.5}$$

(M is any third body, usually N_2 or O_2)

$$NO + O_3 \rightarrow NO_2 + O_2 \tag{6.6}$$

This produces a situation where the concentrations of NO and NO_2 remain constant being controlled by the probability of reaction (6.6). The addition of any other compound that reacts with atomic oxygen in particular will disrupt this pseudo-equilibrium situation. Such is the case in photochemical smog.

The predominant sources of NO are oxidation of NH_3 and high-temperature combustion processes. Both of these are tropospheric sources. Minor sources are found in the stratosphere and thermosphere. In the thermosphere NO is formed by the reaction of oxygen with atomic nitrogen (reaction 6.8)

$$N_2 + h\nu \rightarrow N + N \tag{6.7}$$

$$N + O_2 \rightarrow NO + O \tag{6.8}$$

This is thought to be the principal reaction for the removal of atomic nitrogen from the atmosphere. Reaction (6.2) is also of some importance in the production of stratospheric NO.

Both NO and NO_2 are found in combustion gases, with NO predominating as its formation is favoured at high temperatures. NO is formed in post-flame combustion gases from nitrogen and oxygen in the air used to combust the fuel (reactions 6.9–6.12).

$$CO + OH \rightleftharpoons CO_2 + H \tag{6.9}$$

$$H + O_2 \rightleftharpoons OH + O \tag{6.10}$$

$$O + N_2 \overset{k_1}{\rightleftharpoons} NO + N \tag{6.11}$$

$$O_2 + N \rightleftharpoons NO + O \tag{6.12}$$

Reactions (6.9) and (6.10) occur together in the combustion gases and are largely responsible for the concentration of atomic oxygen. If a steady concentration of nitrogen atoms and combustion under excess air conditions are assumed, it may be shown that the rate of production of NO approximates to eqn (6.13).

$$\frac{d[NO]}{dt} = 2k_1[O][N_2] \tag{6.13}$$

The value of k_1 in reaction (6.11) is $1 \times 10^{11} \exp(-75\,400/RT) \, \text{dm}^3 \, \text{mol}^{-1} \, \text{s}^{-1}$. Application of this to eqn (6.13) shows that the temperature is the most significant factor in the production of NO under normal combustion conditions. The higher the temperature, the greater the production of NO. The production of NO per unit mass of fuel combusted follows the order coal > oil > natural gas, as this is the order of average combustion temperature. In terms of mass emitted, motor vehicles are the most important source of NO as the internal combustion engine operates at a high temperature.

Photochemical smog

The most important atmospheric reactions involving NO and NO_2 occur in the group of reactions known as photochemical smog. A typical photochemical smog occurs in warm sunny weather and it is characterized by a haze, ozone formation, eye irritation, and damage to vegetation. This smog was first recognized in the Los Angeles district hence the following general outline of photochemical smog will refer to the Los Angeles situation.

The geography of the Los Angeles basin is such that the district may be likened to a giant chemical reaction flask. The walls of the 'flask' are made up of the floor of the valley and mountains which occupy three sides of the valley. The fourth side of the valley is open to the ocean but the prevailing wind from the sea effectively closes this wall of the 'flask'. A 'lid' is provided by frequent temperature inversions at 200–500 m. A temperature inversion is found when a layer of warm air overlies cool air adjacent to the ground. This prevents turbulent convective mixing of the air adjacent to the ground with the remainder of the troposphere and thus all gases emitted into this air mass are trapped. A large-scale temperature inversion occurs at the tropopause which effectively separates the troposphere from the stratosphere (Fig. 1.1).

To this reaction 'flask' are added the photochemical smog reagents—exhaust gases from the operation of a large number of motor vehicles. The energy required for the photochemical smog reaction to proceed comes from the solar radiation spectrum. Los Angeles is well-known for its high number of sunlight hours.

The major changes that may be measured in the atmosphere during a photochemical smog episode are illustrated in Fig. 6.1. The sequence of events commences with the injection of hydrocarbons and NO into the atmosphere from the exhaust systems of motor vehicles. With increasing sunlight intensity the concentration of NO decreases, while the concentrations of NO_2 and the aldehydes increase. A decrease in the NO_2 concentration accompanies the appearance of significant O_3 levels which, after midday, show a decrease as do the levels of hydrocarbons and aldehydes. There is no observed increase in NO or hydrocarbon concentrations during the evening traffic peak.

The key reactant in the formation of photochemical smog is NO_2. In conditions where hydrocarbons are absent we have seen that photodissociation sets up a pseudo-equilibrium concentration of NO and NO_2 (reactions 6.4–6.6). In the presence of hydrocarbons this equilibrium is disrupted by a chain reaction initiated by the reaction of atomic oxygen and O_3 with hydrocarbons.

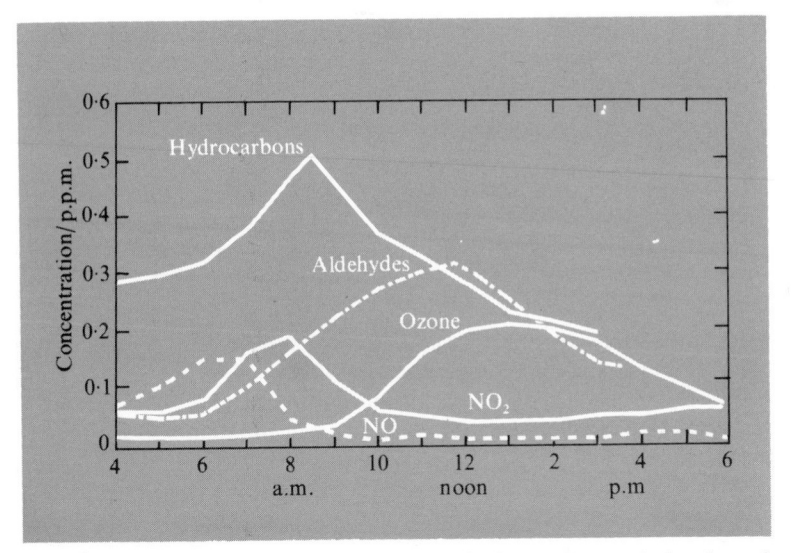

FIG 6.1. Average concentrations of some atmospheric constituents during days of eye irritation in Los Angeles.

The polluted urban atmosphere contains about one hundred different hydrocarbons, the most reactive of which are the olefins. The result of atomic oxygen attack on olefins is the production of two free radicals. In the case of propylene, the first step in the reaction is understood to be the addition of O to the double bond to produce an activated complex (6.14).

$$\begin{array}{c} H \\ \diagdown \\ H_3C \end{array} C=C \begin{array}{c} H \\ \diagup \\ \diagdown H \end{array} + O \ \rightarrow \ \left[\begin{array}{c} H \\ \diagdown \\ H_3C \end{array} C-C \begin{array}{c} H \\ \diagup \\ \diagdown H \end{array} O \right] \tag{6.14}$$

The activated complex may fragment in two different ways (reactions 6.15 and 6.16).

$$\left[\begin{array}{c} H \\ \diagdown \\ H_3C \end{array} C-C \begin{array}{c} H \\ \diagup \\ \diagdown H \end{array} O \right] \ \rightarrow \ H_3C-\overset{\displaystyle H}{\underset{\displaystyle H}{C}}\cdot + \cdot C\overset{\displaystyle O}{\underset{\displaystyle H}{\diagup}} \tag{6.15}$$

$$\left[\begin{array}{c} H \\ \diagdown \\ H_3C \end{array} C-C \begin{array}{c} H \\ \diagup \\ \diagdown H \end{array} O \right] \ \rightarrow \ H_3C\cdot + H_3C-C\overset{O}{\diagdown\!\!\!\diagup} \tag{6.16}$$

Reaction (6.15) is the more likely as it involves less rearrangement of the activated complex than does reaction (6.16). CHO· rapidly forms formaldehyde and CH_3CO· rapidly forms acetaldehyde. Reactions (6.15) and (6.16) are the initiation reactions of a chain reaction (reactions 6.17–6.21).

$$CH_3· + O_2 \rightarrow CH_3O_2· \tag{6.17}$$

$$CH_3O_2· + NO \rightarrow CH_3O· + NO_2 \tag{6.18}$$

$$CH_3O· + O_2 \rightarrow HCHO + NO_2· \tag{6.19}$$

$$HO_2· + NO \rightarrow OH· + NO_2 \tag{6.20}$$

$$C_3H_6 + OH· \rightarrow CH_3CH_2· + H_2O \tag{6.21}$$

It must be emphasized that the chain propagation reactions (6.17) to (6.21) are subject to some conjecture; however they do explain the formation of many of the compounds found in the laboratory studies of photochemical smog. The chain reaction enables a rapid oxidation of NO to NO_2 by alkoxyl (RO·) and peroxoalkyl ($RO_2·$) radicals, thus atomic oxygen and thus O_3 are conserved. This offers some explanation for the observed changes in photo-chemical pollutant gas concentration throughout the day (Fig. 6.1). Other alkenes can also follow this reaction sequence, the overall rate of their oxida-tion varying with their structure.

The initial 'equilibrium' concentrations of NO and NO_2 are controlled by the photolytic NO_2 cycle (reactions 6.4–6.6). As the atmospheric hydro-carbon concentration increases, due to motor vehicle operation, the photo-lytic NO_2 cycle is disrupted and NO is oxidized to NO_2 by the hydrocarbon chain reaction (eqns 6.15–6.21). The low steady-state O_3 concentration found in the photolytic NO_2 cycle thus increases because O_3 is not being used to oxidize NO to NO_2. The hydrocarbon concentration decreases because of its participation in NO oxidation, and the aldehyde concentration increases as it is a product of the hydrocarbon chain oxidation of NO. Of course, the NO concentration decreases and the NO_2 concentration increases as a result of these reactions.

It is now necessary to consider other interactions of the photochemical smog gases in order to explain the further changes in the gas concentrations shown in Fig. 6.1. Hydrocarbon radicals may react with NO_2 to form organic nitrates (reaction 6.22) or peroxoacyl nitrates (reactions 6.23–6.25).

$$RO· + NO_2 + M \rightarrow RNO_3 + M \tag{6.22}$$

$$O + R'H \rightarrow R''· + RCO· \tag{6.23}$$

$$RCO· + O_2 \rightarrow R(CO)O_2· \tag{6.24}$$

$$R(CO)O_2· + NO_2 \rightarrow R(CO)O_2NO_2 \tag{6.25}$$

(M is any third body, usually N_2 or O_2; R, R', and R'' are any hydrocarbon chains.)

Reaction (6.22) is a termination reaction for the hydrocarbon chain oxidation of NO to NO_2. Reaction (6.23) is the first portion of a reaction represented by eqns (6.15) and (6.16). The peroxoacyl radical $R(CO)O_2$ formed by the oxidation of the product of reaction (6.23) is basic to the production of peroxoacyl nitrates (PANs). The PANs are very important in photochemical smog because of their considerable biological reactivity. They cause plant leaf damage at very low concentrations and cause eye and respiratory irritation in humans at concentrations as low as $0.5\ mg\ kg^{-1}$. The best-known PAN is peroxoacetylnitrate ($CH_3C \cdot OO \cdot NO_2$).

The reactions represented by eqns (6.22)–(6.25) thus give some explanation for a decrease in the NO_2 concentration and a further decrease in the hydrocarbon concentration. The formation of the PANs is also explained.

It remains now to consider the reactions between O_3 and the olefinic hydrocarbons. These reactions result initially in the formation of an ozonide intermediate (reaction 6.26).

$$\begin{array}{c}R\\ \diagdown\\ \ \ \ C=C\ \ \ \ + O_3 \rightarrow\\ \diagup \quad \diagdown\\ H \qquad H\end{array} \qquad \left[\begin{array}{c}R \qquad\ \ O \qquad R'\\ \diagdown \diagup \quad \diagdown \diagup\\ C \qquad\qquad C\\ \diagup \diagdown \quad \diagup \diagdown\\ H \quad\ O\!-\!O \quad\ H\end{array}\right] \tag{6.26}$$

where R and R′ are hydrocarbon chains.
The ozonide intermediate may cleave in one of two ways (reactions 6.27 and 6.28)

$$\left[\begin{array}{c}R \qquad\ O \qquad R'\\ \diagdown \diagup \diagup\ \diagup\\ C \qquad\qquad C\\ \diagup \diagup \diagdown \quad \diagdown\\ H \quad O\!-\!O \quad H\end{array}\right] \rightarrow R\dot{C}HO\cdot + R'\dot{C}HOO\cdot \tag{6.27}$$

$$\left[\begin{array}{c}R \quad\ \ O \qquad R'\\ \diagdown \diagdown\ \diagdown \diagup\\ C \qquad\qquad C\\ \diagup \quad \diagdown \diagdown\ \diagdown\\ H \quad O\!-\!O \quad H\end{array}\right] \rightarrow R\dot{C}HOO\cdot + R'\dot{C}HO\cdot \tag{6.28}$$

The biradical products from reactions (6.27) and (6.28) may be written as zwitterions which can undergo a variety of decomposition reactions (such as reactions 6.29–6.31).

$$R'\overset{+}{C}HOO^- \rightarrow R'O\cdot + CHO\cdot \tag{6.29}$$

$$R'\overset{+}{C}HOO^- + NO \rightarrow R'CHO + NO_2 \tag{6.30}$$

$$R'\overset{+}{C}HOO^- + NO_2 \rightarrow R'CHO + NO_3 \tag{6.31}$$

Eqns (6.29)–(6.31) indicate that the zwitterion may decompose to a reactive radical (which may enter the hydrocarbon chain oxidation of NO) and a formyl radical which results ultimately in the formation of formaldehyde. The zwitterion may also oxidize NO and NO_2 being reduced itself to an aldehyde. NO_3 is of some importance in photochemical smog because of its participation in reactions (6.32) and (6.33).

$$NO_3 + NO_2 \rightleftharpoons N_2O_5 \tag{6.32}$$

$$N_2O_5 + H_2O \rightarrow 2HNO_3 \tag{6.33}$$

High concentrations of inorganic nitrate on the atmospheric aerosol present during a photochemical smog may be explained by the production of nitric acid. An alternative source of NO_3 and hence nitric acid is the oxidation of NO_2 with ozone (reaction 6.34). This reaction is very much slower than the ozone oxidation of NO (reaction 6.6).

$$NO_2 + O_3 \rightarrow NO_3 + O_2 \tag{6.34}$$

The afternoon decrease in the concentrations of hydrocarbons and O_3 may now be explained by the reactions considered in the preceding paragraphs. A corresponding oxidation of aldehydes by O_3 offers a similar explanation for the decrease in aldehyde concentration in the afternoon. The evening injection of hydrocarbons and NO by motor vehicles is removed almost immediately by the relatively high concentration of O_3 generated throughout the day. These chemical explanations can only be partial explanations as physical phenomena such as gaseous dispersion and interactions with solid surfaces have not been considered.

In attempting to explain the reactions occurring in photochemical smog, a motor vehicle pollutant gas produced in very high concentration has been omitted. This is CO which was, until recently, thought to be sufficiently unreactive to be ignored. Recent laboratory experiments suggest that this is not the case, so that reactions (6.35)–(6.41) should be considered.

$$NO + NO_2 + H_2O \rightarrow 2HNO_2 \tag{6.35}$$

$$HNO_2 + hv \rightarrow OH\cdot + NO \tag{6.36}$$

$$CO + OH\cdot \xrightarrow{O_2} CO_2 + HO_2\cdot \tag{6.37}$$

$$HO_2\cdot + NO \rightarrow NO_2 + OH\cdot \tag{6.38}$$

$$HO_2\cdot + NO_2 \rightarrow HNO_2 + O_2 \tag{6.39}$$

$$OH\cdot + NO \rightarrow HNO_2 \tag{6.40}$$

$$HO_2\cdot + HO_2\cdot \rightarrow H_2O_2 + O_2 \tag{6.41}$$

These reactions (6.35)–(6.41) provide a further means of oxidizing NO to NO_2 without the participation of atomic oxygen or O_3. They also provide a means for the rapid oxidation of CO to CO_2 which is of some importance in the atmospheric chemistry of CO (see Chapter 7). The production of HNO_2 in reaction (6.35) and its subsequent photolysis in reaction (6.36) to produce hydroxyl radicals is the initiation step of a chain reaction. The propagation steps are reactions (6.37) and (6.38) which provide a means for the oxidation of both CO and NO. Reactions (6.39) and (6.40) are termination steps that produce HNO_2 which may then be used in chain initiation. The production of H_2O_2 has been observed in photochemical smogs and reaction (6.41) provides a mechanism for this. It is also another chain-termination reaction. It is important to note that reactions (6.35) to (6.41) provide a pathway for the oxidation of NO to NO_2 without the participation of hydrocarbons, O_3, or atomic oxygen i.e. the photolytic NO oxidation cycle may be disrupted in the absence of hydrocarbons. The rates of the reactions (6.4)–(6.6) involved in the cyclic photolytic oxidation of NO to NO_2 are, however, much greater than the rates of reactions (6.35)–(6.41), thus the photolytic oxidation is the most important. The pathway through carbon monoxide is only supplementary to it.

Another important pollutant gas that is usually found in photochemical smog situations is SO_2. In Chapter 5 it has been pointed out that SO_2 arises from many combustion processes and thus is usually emitted to the atmosphere in which photochemical smog reactions may be occurring. An analysis of the aerosol associated with photochemical smog indicates a high concentration of sulphate. This is thought to be derived by the oxidation of SO_2 to H_2SO_4 in the presence of some of the constituents of photochemical smog. Possible oxidizing agents include O_3, atomic oxygen, NO_2 and hydrocarbon radicals.

In the absence of photodissociation reactions, NO_2 will oxidize SO_2 only very slowly. When photodissociation of NO_2 occurs in the atmosphere the atomic oxygen so produced can oxidize SO_2 rapidly. However, SO_2 must compete for the atomic oxygen with other molecules, especially O_2 which is present at a concentration of about 10^5 times greater than that of SO_2. Oxidation with atomic oxygen is thus unlikely. The reaction between SO_2 and O_3 in a pure atmosphere is very slow indeed, but in an atmosphere containing hydrocarbons, especially olefins the reaction is very rapid indeed. The oxidation is supposed to be through the agency of peroxo zwitterions (eqn 6.42).

$$RCH\overset{+}{O}O^- + SO_2 \rightarrow RCHO + SO_3 \qquad (6.42)$$

This reaction is analogous to the oxidation of NO by peroxo zwitterions (eqn 6.30). SO_3 is rapidly hydrated with atmospheric water vapour to form a sulphuric acid mist. This mist and the absorption of light by NO_2 in the visible region of the solar spectrum, offer an explanation for the reduction in visibility associated with photochemical smogs.

It is possible to postulate a photochemical smog situation arising from only natural compounds in the atmosphere. We have seen that the photocyclic oxidation of NO is basic to photochemical smog, and that NO can arise from the bacterial oxidation of NH_3. The largest natural source of hydrocarbons in the atmosphere is probably the organic volatiles emanating from plants. These include the terpenes that account for the blue haze seen over large forests. Most natural terpenes have the molecular formula $(C_5H_8)_n$ and are thus unsaturated. It is not surprising therefore, that laboratory experiments have shown that the terpenes, α- and β-pinene, are rapidly oxidized in an atmosphere containing irradiated NO_2.

Damaging effects of NO and NO_2

No cases of human poisoning from NO have been reported because of its relatively low toxicity. Although NO_2 has a higher toxicity to man even its milder effects such as mucous membrane irritation do not occur at atmospheric NO_2 concentrations. With an odour threshold in the range $1-3$ mg kg^{-1} it is seldom detected by the nose in photochemical smog episodes, where the maximum concentration is usually 0.3 mg kg^{-1}.

With vegetation, concentrations of NO_2 in excess of 2 mg kg^{-1} are known to cause leaf damage to sensitive plants. Such concentrations are seldom found in the atmosphere but NO_2 concentrations that are capable of inhibiting photosynthesis (about 0.6 mg kg^{-1}) in some plants are known. The mechanism for this inhibition is, at present, unknown. At lower concentrations plants absorb both NO and NO_2 from the atmosphere without damage. The uptake rate for NO_2 is about twenty times greater than that for NO and its absorption by plants could be significant in the total nitrogen nutrition of the plant.

Damage to materials in the environment by NO and NO_2 is slight. Because it is an oxidizing agent NO_2 may bleach certain dyes e.g. some dyestuffs are substantially faded by treatment for 6 days with 0.7 mg kg^{-1} NO_2 at 50 per cent relative humidity.

Damaging effects of PANs

A typical maximum concentration of PAN in a photochemical smog is 0.03 to 0.04 mg kg^{-1}. Acute toxicity is not found in humans at concentrations of this order but eye irritation occurs after a 12 minute exposure to 0.5 mg kg^{-1} of PAN. Cardio-pulmonary functions are altered at concentrations as low as 0.3 mg kg^{-1} of PAN.

Vegetation effects occur at considerably lower concentrations: sensitive plants may show leaf injuries at PAN concentrations as low as 0.01 mg kg^{-1} and even lower levels may contribute to early leaf fall. Damage usually takes the form of the collapse of young mature cells surrounding the air space in the stomata, hence young leaves are the first affected. Unfortunately, it must again be recorded that the biochemical basis for the damage is largely unknown.

Damaging effects of aldehyde

Maximum atmospheric aldehyde concentrations of up to $0.3\,\text{mg}\,\text{kg}^{-1}$ have been recorded. Formaldehyde in concentrations of this order is known to have deleterious effects on human health. Eye irritation may be experienced at $0.2\,\text{mg}\,\text{kg}^{-1}$ and at as low as $0.13\,\text{mg}\,\text{kg}^{-1}$ dry and sore throats may be experienced. This latter effect may be related to the known inhibition of the respiratory cilia by $0.5\,\text{mg}\,\text{kg}^{-1}$ of formaldehyde. Few data are available on the toxic effects of other organic components of photochemical smog.

7. Carbon monoxide

I F CO_2 is disregarded, the most abundant gaseous pollutant in the atmosphere in the immediate vicinity of most towns is CO. This gas is produced by man by the incomplete combustion of carbonaceous fuels. The largest source of combustion-produced CO is the internal combustion engine. In busy city streets peak concentrations may be over 100 mg kg^{-1} while in tunnels carrying motor vehicles concentrations as high as 300 mg kg^{-1} are found.

Global emissions of CO from combustion sources are estimated at 2.6×10^8 t y^{-1}, almost all of which is emitted in the Northern Hemisphere. For some time it was thought that this was the major source of atmospheric CO, and that natural sources were small and of little significance. Recent evidence now suggests that combustion CO constitutes only about 10 per cent of the total CO source, the remainder arising from two large natural sources—atmospheric reactions and emission from the oceans.

Atmospheric CO sources

Oxidation of CH_4 is probably the most important source of atmospheric CO. As almost all of the CH_4 in the atmosphere is produced by anaerobic decomposition of organic matter, this source of CO is of natural origin.

This oxidation is essentially in two steps—the oxidation of CH_4 by hydroxyl radicals (OH·) to formaldehyde via methyl (CH_3·), methylperoxo (CH_3O_2·) and methoxo (CH_3O·) radicals, followed by the photolysis of formaldehyde.

Methane oxidation is initiated by hydroxyl radicals hence a source of these species is necessary for the reaction to proceed. This is provided by the photochemical decomposition of ozone and the subsequent reaction of atomic oxygen with water vapour (reactions 7.1 and 7.2).

$$O_3 + hv \rightarrow O_2 + O· \tag{7.1}$$

$$·O + H_2O \rightarrow 20H· \tag{7.2}$$

The importance of radiation-initiated reactions in the oxidation of CH_4 explains why this source of CO is often referred to as photochemical.

Oceanic CO sources

A second natural source of CO has been found to be the surface layers of the ocean. Samples of mid-ocean surface waters were found to contain up to ninety times the concentration of CO calculated from standard CO solubility data for the partial pressure of CO in the atmosphere immediately above the ocean surface. This implies that CO is produced in the ocean surface layer and that there is a flux of CO from the ocean to the atmosphere. The degree of supersaturation of CO in the surface waters is increased in the presence of sunlight suggesting two possible mechanisms for the CO production—photochemical oxidation of organic matter or biological oxidation by marine organisms. It appears that the latter source is the most likely.

The size of this marine source has been calculated assuming an average supersaturation of the ocean surface waters of twenty times the equilibrium value. This calculation indicates that the marine source and the combustion source are of approximately the same size in the Northern Hemisphere. The Southern Hemisphere, where combustion CO sources are small, is thus dominated by the natural production of CO.

Destruction of atmospheric CO

An atmospheric residence time for CO of about 0·2 years has been deduced from measurements made on atmospheric ^{14}CO. Cosmic ray neutrons produce ^{14}C atoms from nitrogen in the upper atmosphere. Almost all of these atoms react with oxygen producing ^{14}CO. A knowledge of the rate of production of ^{14}CO by this mechanism, together with a measurement of the abundance of ^{14}CO in the atmosphere, allowed the residence time to be determined.

A relatively rapid rate of CO destruction is required to satisfy such a short residence time. Measurements of the vertical concentration profile of CO show an approximately constant CO concentration throughout the troposphere with a sharp drop above the tropopause in the stratosphere (see Fig. 7.1).

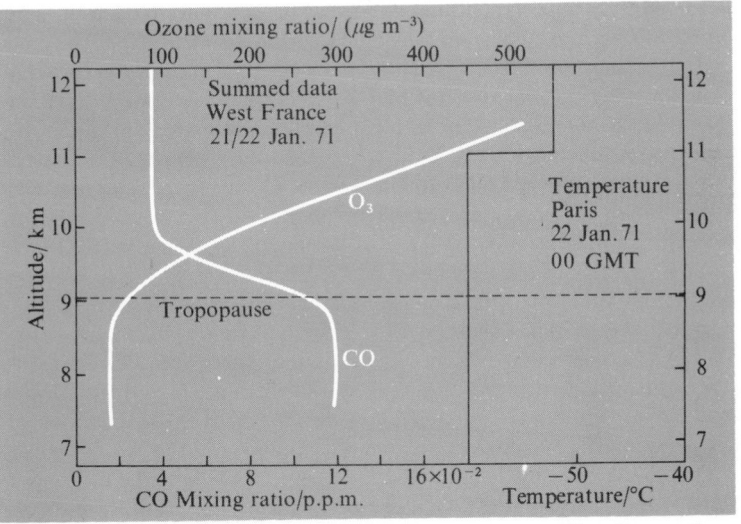

FIG 7.1. CO and O_3 concentration profiles from ascents over West France (from Seiler and Warneck 1972).

Following the rapid decrease a steady, but lower concentration is found through the remainder of the stratosphere. A very active destruction of CO thus occurs in the lower stratosphere.

The reaction thought most likely to account for the destruction of CO in the lower stratosphere is represented by eqn (7.3)

$$OH + CO \rightarrow CO_2 + H \tag{7.3}$$

The hydroxyl radicals arise largely from reactions (7.1) and (7.2). The constant CO concentration higher in the stratosphere is probably due to a source in this region whose rate of CO production is the same as the rate of CO oxidation in reaction (7.3). This source is probably the oxidation of methane which has diffused from the troposphere.

In the troposphere oxidation reaction (7.3) is also possible, especially in regions where photochemical smog is being actively formed (see reactions 6.34–6.40). Because much CO is emitted with other motor-vehicle exhaust gases, a considerable portion of the oxidation of exhaust-CO may occur rapidly. Within the troposphere as a whole, turbulent mixing with naturally-produced CO would make any observation of this oxidation very difficult.

At the surface of the earth CO may be destroyed by biological mechanisms. Soil fungi are known to absorb CO at an average rate of $2 \mu g \, s^{-1}$ per square metre of soil. The total soil surface of the United States is sufficient to absorb three times the total combustion CO produced in the world. As combustion CO is produced close to the surface of the earth, soil fungal activity must be regarded as an important means of reducing the concentration of CO emitted to the atmosphere by man's activity.

Higher plants also have an ability to absorb CO, but the absorption rate is so low that it can be detected only by sensitive radiochemical techniques. The amount of CO taken up by plant leaves varies with plant species from zero up to about $2 \mu g \, s^{-1}$ per m^2 of ground area. This suggests that plant leaves have an ability to absorb atmospheric CO comparable to that of soil fungi. Biological uptake of CO is thus very important in reducing tropospheric CO concentrations at the earth's surface.

Some aspects of the mechanism of CO destruction by higher plants are known. In the light, the predominant mechanism appears to be reduction of CO probably to 5-formyl-tetrahydrofolic acid which is a well-known biological carrier of groups containing one carbon atom. The reduced CO is fixed as the amino acid serine and thence enters normal protein and carbohydrate metabolism. In the dark, almost all of the CO taken up by plant leaves is oxidized to CO_2 and released again to the atmosphere. The rate of uptake of CO is the same in the light and dark and, as both light and dark processes lead to the destruction of CO, it can be taken that the plant leaves operate throughout the whole day destroying atmospheric CO at an approximately constant rate.

Toxicity of CO

The only known damaging effects of CO at the concentrations found in the atmosphere are related to animal respiratory systems based on haemoglobin as an oxygen-transporter. Haemoglobin is made up of four haem molecules, which are complexes of Fe^{II}, bound to one molecule of the protein globin. The haem molecule has the structure shown in Fig. 7.2. Magnetic

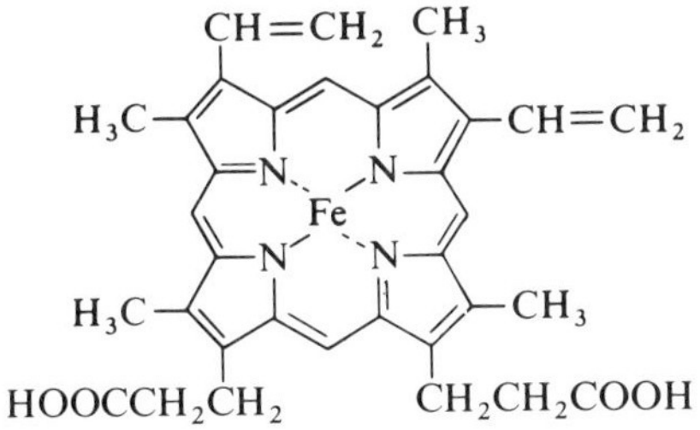

FIG. 7.2. Structure of the haem molecule.

susceptibility measurements have shown that Fe^{II} in haem is in a d^6 octahedral configuration with four unpaired electrons. The four nitrogen atoms of the organic chelate occupy four of the octahedral coordination positions in roughly the same plane. One of the positions perpendicular to the plane is occupied by coordination to the globin molecule, while the other position is available for coordination of (usually) oxygen gas. When oxygen gas is co-ordinated the molecule is known as oxyhaemoglobin.

The affinity of CO for the coordination position usually occupied by oxygen is about 200 times greater than the affinity of oxygen for the position. A relatively low partial pressure of CO is thus able to displace a considerable amount of oxygen from oxyhaemoglobin (HbO_2) to form the CO–haemoglobin complex known as carboxyhaemoglobin (HbCO), as shown in eqn (7.4).

$$HbO_2 + CO \rightleftharpoons HbCO + O_2 \qquad (7.4)$$

The transport of oxygen from the lungs to the tissues is thus impaired.

Reaction (7.4) is reversible, thus when the partial pressure of CO in the lungs is reduced, the reaction moves to the left and CO is released. The use of pure oxygen in the treatment of CO poisoning is based on this fundamental chemical principal.

Many individuals suffer from oxygen transport problems when the carboxyhaemoglobin content of their blood reaches 5 per cent. This is attained after exposure to 30 mg kg^{-1} CO for four hours or 120 mg kg^{-1} CO for one hour. It is of interest to note that cigarette smokers commonly have a carboxyhaemoglobin content in their blood of 5 to 10 per cent.

8. Minor pollutant gases

THE gases to be considered in this chapter are minor only in that their effects on the environment are overshadowed by the effects of the compounds discussed in earlier chapters. The compounds discussed are only a few of many that could have been chosen, however they have a common property in that they are all compounds that exist naturally in the atmosphere and have a man-made source.

The halogens

Fluorine is present in the unpolluted atmosphere in very low concentration in both gaseous and particulate form. It is most likely derived from sea-water aerosols formed by the bursting of bubbles in the oceans, especially in 'white caps'. The origin of the gaseous fraction is not clear. Oxidation of fluoride by ozone in the droplet phase, as has been suggested by some workers, is unlikely as the standard electrode potentials of O_3 and F^- are 2·07 and 2·87 V respectively.

Heavy industries contribute significant concentrations of fluorine to the atmosphere. The phosphate fertilizer industry releases fluorine largely as H_2SiF_6 from the action of H_2SO_4 on the fluorides in phosphate rock. Cryolite (Na_3AlF_6) is used as a flux with bauxite (Al_2O_3) in the aluminium industry. Fluorine released from this source appears as SiF_4, HF, or Na_3AlF_6.

Fluorine concentrations (gaseous plus particulate) can reach 5 μg m^{-3} in areas around industrial sources. These concentrations, as such, cause no immediate damage to plants or animals but fluorides are known to act as cumulative poisons to plants. Gaseous fluorides enter the leaves through the stomata and are transported to the edges of the leaves where they accumulate. With leaf concentrations in the range 50 to 200 mg kg^{-1} sensitive plants show leaf damage, however resistant species can tolerate up to 500 mg kg^{-1} of fluoride before being damaged.

Fluorine in the atmosphere does not have a direct effect on animals but grazing animals are indirectly affected. When atmospheric fluorine is accumulated in grass and forage crops the concentrations may be sufficiently high to cause fluorosis, which is characterized by bone damage. Safe fluoride levels in animal feed are about 50 mg kg^{-1}. Concentrations of this order are found in the vegetation about some large fluorine-emitting sources.

Chlorine also occurs in both gaseous and particulate forms in the unpolluted atmosphere. The source is largely sea-water droplets which may carry the chloride ion in solution or as a solid salt following evaporation. Both solid and gaseous chlorine in the atmosphere appear in the concentration range

0.5–5 μg m^{-3}. The nature of the gaseous form has not been determined but it is thought to be HCl and Cl_2. The former is thought to be formed in chloride-containing droplets by the action of H_2SO_4 on chlorides, (reaction 8.1), the H_2SO_4 being derived from the oxidation of SO_2.

$$2Cl^- + H_2SO_4 \rightarrow 2HCl + SO_4^{2-} \qquad (8.1)$$

Releases of free chlorine from industry are very unusual but a widespread pollution source exists in the form of motor vehicle emissions. These contain lead halide aerosols formed from 'anti-knock' compounds in the petrol. Photochemical decomposition of the halide produces chlorine atoms. These may participate in NO oxidation as shown in reactions (8.2)–(8.8).

$$Cl + O_2 + M \rightarrow ClO_2 + M \qquad (8.2)$$

$$ClO_2 + NO \rightarrow ClO + NO_2 \qquad (8.3)$$

$$ClO + NO \rightarrow Cl + NO_2 \qquad (8.4)$$

$$2ClO \rightarrow Cl + ClO_2 \qquad (8.5)$$

$$Cl + ClO_2 \rightarrow Cl_2 + O_2 \qquad (8.6)$$

$$2Cl + M \rightarrow Cl_2 + M \qquad (8.7)$$

$$Cl_2 + h\nu \rightarrow 2Cl \qquad (8.8)$$

Reactions (8.2)–(8.5) are propagation steps of a chain reaction oxidizing NO to NO_2 while reactions (8.6) and (8.7) are termination reactions. Reaction (8.8) is an initiation reaction that acts as an alternative to lead halide photo-decomposition.

An alternative man-made source of HCl in particular, is the combustion of chlorinated organic polymers, e.g. polyvinylchloride. The combustion of solid waste materials at rubbish dumps is a frequent source of this type.

Damage to human health by chlorine in the atmosphere has only been found in the high concentrations due to accidental spills. Plants, however, may be damaged by chlorine gas at concentrations as low as 0.2 mg kg^{-1}. At lower concentrations (about 0.1 mg kg^{-1}) this gas can cause partial closure of the leaf stomata. Even with this effect, which will reduce the uptake of chlorine by the plant leaf, it can be shown that vegetation uptake is an important means of reducing the concentration of atmospheric chlorine. Non-damaging concentrations of HF and HCl are similarly reduced when the atmosphere containing them passes over plant leaf surfaces.

Bromine has received little attention as a gas in the atmosphere. Like the other halogens its natural source is thought to be sea-salt droplets. It exists in the atmosphere in two forms—gaseous and particulate—with the former being about five times the concentration of the latter. The reasons for this observation are not yet understood.

The most important pollutant source of bromine in the atmosphere is the combustion of petrol containing lead 'anti-knock' compounds. The products of combustion include $PbClBr$, $NH_4Cl \cdot 2PbClBr$, and $2NH_4Cl \cdot PbClBr$. Photochemical decomposition of these lead halides result in the release of gaseous bromine. Some of this remains in the gas phase while the remainder may react with, or become absorbed on, solid or liquid atmospheric aerosols. Damage to vegetation by atmospheric concentrations of bromine has not been reported.

The concentrations of iodine found in the atmosphere fall in the range 0.01 to $10 \, \mu g \, m^{-3}$ with an average of about $0.1 \, \mu g \, m^{-3}$. As with the other halogens, both particulate and gaseous forms are found and these are thought to originate from the ocean. The concentration of gaseous iodine is several times greater than that of the particulate form. The gaseous iodine is generated at the surface of the ocean and is probably released directly from the ocean. Several mechanisms for the production of gaseous iodine have been proposed, including a photochemical reaction (eqn 8.9) and a chemical oxidation (eqn 8.10).

$$2I^- + \tfrac{1}{2}O_2 + H_2O + h\nu \rightarrow I_2 + 2OH^- \tag{8.9}$$

$$2I^- + \tfrac{1}{2}O_2 + 2H^+ \rightarrow I_2 + H_2O \tag{8.10}$$

Combustion of fossil fuels, which can contain up to $5 \, mg \, kg^{-1}$ of iodide, is a source of man-made iodine in the atmosphere. A radio-active isotope of iodine ^{131}I is released to the atmosphere as a result of nuclear fission. As ^{131}I is accumulated in the thyroid it has a high radiological toxicity. For this reason the chemistry of ^{131}I-iodine species in the atmosphere has been subjected to considerable study. The results obtained apply equally well to the naturally-occurring isotopes of iodine.

[131]I released to the atmosphere may be partially absorbed by the particulate matter in the atmosphere, the extent of the absorption varying with the concentration of particulate matter. In country districts with average smoke concentrations about 15 per cent is absorbed, while in air highly polluted with smoke up to 70 per cent may be absorbed. The gaseous forms of ^{131}I-iodine include I_2 gas and aliphatic iodides the most abundant of which is CH_3I. This latter is absorbed by particulate matter and solid surfaces to a much lesser extent than is I_2. Vegetation and soil have been found to be important surfaces for the removal of ^{131}I-iodine compounds from the atmosphere.

Hydrocarbons

In Chapter 6 the most important reactions of the hydrocarbon gases emitted to the atmosphere were considered in the discussion on photochemical smog. In this section the nature and origin of pollutant hydrocarbon gases will be considered together with the atmospheric chemistry of some of the most

TABLE 8.1

Average concentration of some hydrocarbons in urban
atmospheres (from Grob and Altshuller et al.)

Hydrocarbon	Average concentration mg kg^{-1}
Methane	2·0
Ethane	0·05
Ethylene	0·03
Acetylene	0·03
n-Butane	0·03
Isopentane	0·02
Propane	0·02
Toluene	0·02
n-Pentane	0·02
m-Xylene	0·02
Isobutane	0·02
Propylene	0·01
Butenes	0·01

abundant compounds. In Table 8.1 are listed some of the hydrocarbons found in the urban atmosphere.

Methane appears to be present in an anomalously high concentration, but this is because of the considerable natural production of methane due to anaerobic bacterial decomposition of organic matter and emission from geothermal areas, coalfields, natural gas, and petroleum wells. This gives rise to a natural atmospheric methane concentration of about 1·4 mg kg^{-1}, and an estimated average life of methane in the atmosphere of up to 20 years.

About one half of the methane emission can be related to human activities which increase the turnover of organic matter at the surface of the earth. This could possibly be defined as pollutant methane. The production of methane in combustion processes, especially motor vehicle engines, is a smaller but more obvious pollutant source.

The ocean is almost in equilibrium with the partial pressure of methane in the atmosphere. There is a suggestion that the activity of micro-organisms in the surface layers of the ocean may be the origin of a small source of methane. The concentration of methane in the troposphere is virtually constant, but it falls off rapidly in the lower stratosphere. Oxidation by hydroxyl radicals to CO and water vapour is thought to cause this effect. Some concern has been expressed recently regarding the possible increased water vapour concentration in the stratosphere due to the oxidation of increasing methane emissions from the earth's surface. The water vapour may destroy the ozone equilibrium (see Chapter 4) by the production of hydroxyl radicals (reactions 7.1 and 7.2). A depletion in the stratospheric ozone concentration would give rise to an increase in the flux of ultraviolet light reaching the surface of the earth.

Of the hydrocarbons listed in Table 8.1 methane, ethane, propane, and to a lesser extent isobutane are derived largely from sources other than motor vehicle exhausts. Toluene and *m*-xylene have an exhaust origin as well as an industrial origin because of their importance as solvents in industry. The remainder are largely of motor vehicle exhaust origin. Methane, ethane, propane, and acetylene are of low reactivity in photochemical smog, while the olefins are of high activity.

Ethylene is the only hydrocarbon listed in Table 8.1 that is capable of damage to biological systems without further reaction and at atmospheric concentrations. The other hydrocarbons, because of their participation in reactions producing O_3, PAN etc. have an indirect damaging effect on biological systems. Ethylene is a plant growth hormone produced naturally at very low concentrations by many plants. Concentrations of 0.005 mg kg^{-1} of ethylene in the atmosphere can cause leaf damage to very sensitive plants, while less sensitive plants such as tomatoes can show growth retardation after exposure to 0.05 mg kg^{-1} for several weeks.

Ammonia and ammonium sulphate

Ammonia is released to the atmosphere in gaseous form. More than 80 per cent of the release is from natural sources especially the hydrolysis of urea from animal urine. No more than 20 per cent is released from the combustion of the nitrogen fraction of coal or from industrial processes using ammonia. Once in the atmosphere ammonia may remain as a gas or be found as the ammonium ion, largely in the form of ammonium sulphate. This latter is found in a crystalline form at relative humidities below 81 per cent and in the form of droplets at above 81 per cent relative humidity. The concentration of total ammonia (gas plus (NH_4SO_4)) in the atmosphere is about 5 μg m^{-3} and this remains relatively constant throughout the year.

The source of sulphate in atmospheric ammonium sulphate is largely SO_2. In Chapter 5 we have seen that SO_2 oxidation in the atmosphere may take place in the gas phase or the droplet phase. It is most likely that ammonium sulphate is formed in the droplet phase. When evaporation occurs ammonium sulphate particles are left suspended in the atmosphere. These are very effective condensation nuclei i.e. particles on which water vapour may attain the liquid state. This property enables atmospheric ammonium sulphate to exert an effect on visibility as very small water droplets markedly reduce visibility by light scattering. Calculations of this effect for a real situation show that when no condensation on ammonium sulphate particles has occurred (below 81 per cent relative humidity) the average visual range is 35 km while at higher humidity (90 per cent) when some condensation of water has taken place the average visual range is reduced to 24 km.

9. Indoor pollution

I N the foregoing chapters the impression may have been gained that exposure to air pollutants is an out-of-doors phenomenon. No real consideration has been given to indoor effects and this reflects the state of knowledge in this area. Very few data are available on the sources of indoor pollutants and of the reactions of these pollutants in a confined volume. In this chapter an attempt will be made to review the results of recent research in this area.

Indoor pollutant concentration

Studies with SO_2 have shown that the indoor concentration of SO_2 can be as low as 20 per cent of that prevailing outdoors. In one experiment a pulse of SO_2-polluted air was admitted to a room through a window. The SO_2 concentration in the room was continuously monitored and it was found that the concentration decreased with a first-order half-life of 40 to 60 minutes. Another experiment, carried out in a large test room coated with polyurethane resin, gave a half-life of 10 hours for the decrease in SO_2 concentration. Obviously the nature of the surfaces in the room has a significant effect on the behaviour of SO_2 in that room.

Data obtained for CO concentrations show that no significant decrease in CO concentration is found indoors compared with outdoors. This observation is consistent with the low capacity for absorption shown by CO. It also indicates the ease with which external air exchanges with indoor air.

Some measurements have been made of the indoor concentration of the solid atmospheric aerosol. A reduction of outdoor concentration to as low as 20 per cent is found indoors, the reduction being greatest in air-conditioned buildings. Concentrations ranged from 30 to 80 $\mu g\ m^{-3}$ of solid particulate material indoors in city buildings, both commercial and residential. Outdoor concentrations ranged from 80 to 190 $\mu g\ m^{-3}$. Analyses for Pb in the solid aerosol in buildings show that it follows the same concentration reduction as the total solid aerosol. Indoor concentrations in city buildings ranged from 0·2 to 2 $\mu g\ m^{-3}$ of Pb.

Interaction of pollutants with indoor materials

The reduction in the concentration of the solid aerosol indoors as compared with that outdoors, can be explained by the considerable capacity of solid aerosols to adhere to surfaces. This property has previously been mentioned in Chapter 2.

The comparable effect with SO_2 is due to its great capacity for sorption on surfaces. In recent years the capacity of many indoor materials to sorb SO_2 has been investigated. Some results are summarized in Table 9.1 where the

TABLE 9.1

Summary of average deposition velocities for SO₂ onto indoor surfaces

Surface	Velocity of deposition (cm sec^{-1})	Area of surface in a typical house (m^2)	Mass of SO_2 deposited in 1 hour from a pulse of 500 μgm^{-3} SO_2 (μgm)	Relative uptake based on unity for waxed linoleum
Wallpaper, PVC	0·003	100	5400	14
Wallpaper, cellulose	0·015	140	38 000	100
Timber, hardwood	0·048	—	—	—
Timber, softwood	0·024	—	—	—
Leather, upholstery	0.16	5	14 000	37
Linoleum, unwaxed	0·003	—	—	—
Linoleum, waxed	0·0006	35	380	1
Carpet, wool	0·021	50	19 000	50
Carpet, nylon	0·007	15	1900	5
Paint, gloss	0·020	60	22 000	58
Paint, emulsion	0·24	100	430 000	1130
Furnishing fabric,				
cotton	0·22	10	40 000	105
wool	0·29	10	52 000	137
artificial	0·026	20	9400	25

capacity of a given material to sorb SO_2 has been expressed as a velocity of deposition, v_g (see Chapter 5).

Also in Table 9.1 are the arbitrarily-selected surface areas of various materials in a 'typical house'. Use has been made of this parameter to show the likely distribution on indoor surfaces of SO_2 released within a house. The high v_g value for SO_2 deposition onto emulsion paint, together with the relatively high surface area of this material, makes it the most likely surface for the sorption of SO_2. In the 'typical house' these emulsion-painted surfaces have been assumed to be ceilings. It is of interest to note that about 40 per cent of the sulphur-35 SO_2 accidentally released in a laboratory was found on an emulsion-painted ceiling.

Upholstery leather also has a high capacity to sorb SO_2 but, as its surface area in a 'typical house' is small, it accounts for only a small portion of total SO_2 sorbed indoors. In the 'typical house' the second most important surface for the removal of SO_2 is furnishing fabric which accounts for about 15 per cent of the gaseous deposition. The natural fibres have a SO_2 deposition velocity of an order of magnitude greater than that of the artificial fibres. It is known that cellulosic fabrics undergo chemical changes, in the presence of atmospheric concentrations of SO_2, that lead to loss of strength probably because of H_2SO_4 hydrolysis.

Measurements of indoor and outdoor SO_2 concentrations in some Dutch homes have shown that indoor surfaces lose their ability to sorb SO_2 as time goes on. This was particularly obvious with painted surfaces and lime-washed surfaces. Indoor SO_2 concentrations in houses where the major absorbing surfaces were old thus approached the SO_2 concentrations prevailing outside.

Indoor pollution sources

Most pollutants measured within a building have a source outside that building, but some measurements have indicated that certain pollutants may arise from indoor sources. A particular example of such a source is a furnace operated to supply heating for a building. A poorly-maintained furnace may leak into the indoor air gases, such as CO and SO_2, as well as solid particulate material, depending upon the type of fuel used. In cases of this type the concentration gradient for CO and SO_2 may be from the building to the outside air.

Other indoor sources include gas stoves and gas heaters. These may give rise to their constituent organic gases (and CO in the case of coal gas) as well as a finite concentration of SO_2 from the combustion of sulphur-containing odourizers in the gas. Cooking also releases materials into the atmosphere, especially mists of cooking fats and oils. These are generally of a very small radius and remain in the air for some time before sedimentation. Their deleterious effect on clean surfaces about the cooking area is well known.

Cigarette smoking provides an indoor source of pollutants including CO and organic particulate material. Indoor sources of this type produce higher concentrations of pollutants than outdoors because of the limited volume of indoor air into which they are released.

Many people spend the major portion of the day indoors hence studies on the effects of air pollutants on human health should take into account the varying concentrations of pollutants to which people are exposed during the whole day.

References

Chapter 1

CHANDLER, T. J. (1967). *The air around us*, Aldus Books, London.

POGOSYAN, KH. P. (1965). *The air envelope of the earth*, Israel Program for Scientific Translations, Jerusalem.

Chapter 2

JUNGE, C. E. (1963). *Air chemistry and radioactivity* ch. 2. Academic Press, New York.

MILLER, M. S., FRIEDLANDER, S. K., and HIDY, G. M. (1972). *J. Colloid Interface Sci.* **39**, 165–176.

NOVAKOV, T., MUELLER, P. K., ALCOCER, A. E., and OTVOS, J. W. (1972). *J. Colloid Interface Sci.* **39**, 225–234.

PEIRSON, D. H., CAWSE, P. A., SALMON, L., and CAMBRAY, R. S. (1973). *Nature* **241**, 252–256.

STERN, A. C. (1968). *Air pollution* 2nd edn. vol. 1. Academic Press, New York.

Chapter 3

ATTIWILL, P. M. (1971). *Environ. Pollut.* **1**, 249.

DYER, A. J. (1972). *Proceedings of International Clean Air Conference, Melbourne, Australia, May* 1972, pp. 12–16.

GARRATT, J. R. and PEARMAN, G. I. (1972). *Proceedings of International Clean Air Conference, Melbourne, Australia, May* 1972, pp. 17–22.

HOOVER, T. E. and BERKSHIRE, D. C. (1969). *J. Geophys. Res.* **74**, 456.

JUNGE, C. E. (1963). *Air chemistry and radioactivity*, pp. 4–36. Academic Press, New York.

SAWYER, J. S. (1972). *Nature* **239**, 23.

Chapter 4

HILL, A. C. and LITTLEFIELD, N. (1969). *Environ. Sci. Technol.* **3**, 52–6.

JUNGE, C. E. (1963). *Air chemistry and radioactivity*, pp. 37–59. Academic Press, New York.

Chapter 5

EVANS, U. R. (1972). 'Mechanism of rusting under different conditions', *Br. Corros. J.* **7**, 10–14.

LISS, P. S. (1971). 'Exchange of SO_2 between the atmosphere and natural waters', *Nature* **233**, 327–9.

ROBINSON, E. and ROBBINS, R. C. (1970). 'Gaseous sulphur pollutants from urban and natural sources', *J. Air Pollut. Contr. Assn.* **20**, 233–5.

STERN, A. C. (ed.) (1968). *Air pollution* 2nd edn., vol. 1. Academic Press, New York.

URONE, P. and SCHROEDER, W. H. (1969). 'SO_2 in the Atmosphere', *Environ. Sci. and Technol.* **3**, 436–45.

Chapter 6

ALTSHULLER, A. P. and BUFALINI, J. J. (1971). *Environ. Sci. and Technol.* **5**, 39–64.

HECHT, T. A. and SEINFELD, J. H. (1972). *Environ. Sci. and Technol.* **6**, 47–57.

STERN, A. C. (1968). *Air pollution* 2nd edn., vol. 1 Academic Press, New York.

Chapter 7

BIDWELL, R. G. S. and FRASER, D. E. (1972). *Can. J. Bot.* **50**, 1435–9.
INMAN, R. E., INGERSOLL, R. B., and LEVY, E. A. (1971). *Science* **72**, 1229–31
JUNGE, C., SEILER, W., and WARNECK, P. (1971). *J. Geophys. Res.* **76**, 2866–79
WOFSY, S. C., McCONNELL, J. C., and McELROY, M. B. (1972). *J. Geophys. Res.* **77**, 4477–93.

Chapter 9

WILSON, M. J. G. (1968). *Proc. R. Soc. A.* **300**, 215–21.
YOCOM, J. E., CLINK, W. L., and COTE, W. A. (1971). *J. Air Pollut. Contr. Assn.* **21**, 251–9.

Index

Periodic Table of the Elements

IA	IIA	IIIA	IVA	VA	VIA	VIIA	VIII			IB	IIB	IIIB	IVB	VB	VIB	VIIB	O
1H 1·008																	2He 4·003
3Li 6·941	4Be 9·012											5B 10·81	6C 12·01	7N 14·01	8O 16·00	9F 19·00	10Ne 20·18
11Na 22·99	12Mg 24·31											13Al 26·98	14Si 28·09	15P 30·97	16S 32·06	17Cl 35·45	18Ar 39·95
19K 39·10	20Ca 40·08	21Sc 44·96	22Ti 47·90	23V 50·94	24Cr 52·00	25Mn 54·94	26Fe 55·85	27Co 58·93	28Ni 58·71	29Cu 63·55	30Zn 65·37	31Ga 69·72	32Ge 72·59	33As 74·92	34Se 78·96	35Br 79·90	36Kr 83·80
37Rb 85·47	38Sr 87·62	39Y 88·91	40Zr 91·22	41Nb 92·91	42Mo 95·94	43Tc 98·91	44Ru 101·1	45Rh 102·9	46Pd 106·4	47Ag 107·9	48Cd 112·4	49In 114·8	50Sn 118·7	51Sb 121·8	52Te 127·6	53I 126·9	54Xe 131·3
55Cs 132·9	56Ba 137·3	57La 138·9	72Hf 178·5	73Ta 180·9	74W 183·9	75Re 186·2	76Os 190·2	77Ir 192·2	78Pt 195·1	79Au 197·0	80Hg 200·6	81Tl 204·4	82Pb 207·2	83Bi 209·0	84Po (210)	85At (210)	86Rn (222)
87Fr (223)	88Ra 226·0	89Ac (227)															

Lanthanides	57La 138·9	58Ce 140·1	59Pr 140·9	60Nd 144·2	61Pm (147)	62Sm 150·4	63Eu 152·0	64Gd 157·3	65Tb 158·9	66Dy 162·5	67Ho 164·9	68Er 167·3	69Tm 168·9	70Yb 173·0	71Lu 175·0
Actinides	89Ac (227)	90Th 232·0	91Pa 231·0	92U 238·0	93Np 237·0	94Pu (242)	95Am (243)	96Cm (247)	97Bk (247)	98Cf (251)	99Es (253)	100Fm (256)	101Md (254)	102No (254)	103Lw (257)

SI units

Physical quantity	Old unit	Value in SI units
energy	calorie (thermochemical)	4·184 J (joule)
	*electronvolt—eV	$1·602 \times 10^{-19}$ J
	*electronvolt per molecule	96·48 kJ mol^{-1}
	erg	10^{-7} J
	*wave number—cm^{-1}	$1·986 \times 10^{-23}$ J
entropy (S)	eu = cal g^{-1} °C^{-1}	4184 J kg^{-1} K^{-1}
force	dyne	10^{-5} N (newton)
pressure (P)	atmosphere	$1·013 \times 10^{5}$ Pa (pascal), or N m^{-2}
	torr = mmHg	133·3 Pa
dipole moment (μ)	debye—D	$3·334 \times 10^{-30}$ C m
magnetic flux density (H)	*gauss—G	10^{-4} T (tesla)
frequency (ν)	cycle per second	1 Hz (hertz)
relative permittivity (ε)	dielectric constant	1
temperature (T)	*°C and °K	1 K (kelvin); 0 °C = 273·2 K

(* indicates permitted non-SI unit)

Multiples of the base units are illustrated by length

fraction	10^9	10^6	10^3	1	(10^{-2})	10^{-3}	10^{-6}	10^{-9}	(10^{-10})	10^{-12}
prefix	giga-	mega-	kilo-	metre	(centi-)	milli-	micro-	nano-	(*ångstrom)	pico-
unit	Gm	Mm	km	m	(cm)	mm	μm	nm	(*Å)	pm

The fundamental constants

Avogadro constant	L or N_A	$6·022 \times 10^{23}$ mol^{-1}
Bohr magneton	μ_B	$9·274 \times 10^{-24}$ J T^{-1}
Bohr radius	a_0	$5·292 \times 10^{-11}$ m
Boltzmann constant	k	$1·381 \times 10^{-23}$ J K^{-1}
charge of a proton (charge of an electron = $-e$)	e	$1·602 \times 10^{-19}$ C
Faraday constant	F	$9·649 \times 10^{4}$ C mol^{-1}
gas constant	R	$8·314$ J K^{-1} mol^{-1}
nuclear magneton	μ_N	$5·051 \times 10^{-27}$ J T^{-1}
permeability of a vacuum	μ_0	$4\pi \times 10^{-7}$ H m^{-1} or N A^{-2}
permittivity of a vacuum	ε_0	$8·854 \times 10^{-12}$ F m^{-1}
Planck constant	h	$6·626 \times 10^{-34}$ J s
(Planck constant)/2π	$\hbar$	$1·055 \times 10^{-34}$ J s
rest mass of electron	m_e	$9·110 \times 10^{-31}$ kg
rest mass of proton	m_p	$1·673 \times 10^{-27}$ kg
speed of light in a vacuum	c	$2·998 \times 10^{8}$ m s^{-1}

$\ln 10 = 2·303$ $\ln x = 2·303 \lg x$ $\lg e = 0·4343$ $\pi = 3·142$
$R \ln 10 = 19·14$ J K^{-1} mol^{-1} $RTF^{-1} \ln 10 = 59·16$ mV at 298·2 K

THE INCONVENIENT CORPSE

A top London villain like Georgie Parks had no business getting himself murdered in the sleepy village of Cookstone. There was no proper hospital, only a small police unit, and a general lack of facilities. It was nothing short of inconsiderate of Georgie, in local Superintendent Michael Brent's opinion.

Chief Inspector 'Actor' Sam Hudson of Scotland Yard's Serious Crimes Squad took a different view. The murder provided him with an opportunity to stir up trouble between rival gangs, and he seized it. Notorious Blackie Webber was quickly involved, and so was stolen jewellery, plus a mysterious connection with the Arena, an expensive gay bar.

But, if predatory Nancy Parfitt had not been frustrated, the connection with Cookstone might never have emerged.

THE INCONVENIENT CORPSE

Philip Daniels

First published 1982
by
Robert Hale Ltd

This edition 2008 by BBC Audiobooks Ltd
published by arrangement with
the author's estate

ISBN 978 1 405 68583 2

British Library Cataloguing in Publication Data available

Printed and bound in Great Britain by
Antony Rowe Ltd., Chippenham, Wiltshire

Prologue

The high, dark clouds obscuring the moon moved gently away, and it was as though someone had switched on a light in the quiet lane. P.C. Tom Hibbert pushed easily against the bicycle pedals and looked at his watch. Almost two-thirty. In half an hour he would be able to make his routine stop at the all-night petrol station. It would be warm there, and a mug of steaming coffee would be quickly forthcoming. Personally, he was of the view that the owners ought to be prosecuted, under some heading or other, for leaving only one person on duty during the night hours. It was an open invitation to the grab-artists, and they'd paid the place a visit twice in the past year. He was always hoping someone might try it on while he was there. That would make a welcome break for himself, and a nice surprise for the intruders.

A few yards ahead, a rabbit darted suddenly across, followed at once by another. Hibbert grinned. Those were about the only intruders he was likely to come up against on this shift. Fat chance of any – hullo, what was that? In front of the hedge, through which the rabbits had disappeared, was a dark, indistinguishable heap of something. Something which had no business to be there. Angling across the lane, he drew closer. A pile of old clothes? No, wait a minute. A tramp, a sleeping tramp, that's what it was. Well, this would brighten things up a bit.

He dismounted from the bicycle, laying it carefully sideways on the grass, and approached the prostrate figure. H'm. Too well-dressed for a tramp. Chap had had too much to drink, or perhaps he was ill.

Hibbert bent down, putting a hand on the man's shoulder.

"Come along now. Wakey, wakey."

The man was curled up, face down in the thick grass, and he made no response. The next shake was less gentle.

"Wake up. I'm a police officer."

That usually managed to penetrate the fumes, but not this time. Perhaps the man was in a coma, or unconscious. Grasping him firmly, the policeman turned the motionless figure over. Sightless eyes stared up at him, and he felt a sudden chill. The man's jacket had fallen open, and there was a mass of wet, sticky blood over his shirt. There would be no point in checking his pulse or heart beat. This man was dead. Murdered, by the look of it.

The night was going to be busy after all.

One

Superintendent Michael Brent stood at the window, staring out at the autumn sunshine. This case had got off to a bad start, so far as he was concerned, and what he must do was to keep reminding himself that it really wasn't anyone's fault. Losing his famous temper with the local coppers would achieve nothing, and would in any case be quite unfair.

The point was, he had not seen the body in situ. The reasons were perfectly valid, but that did not lessen his annoyance. He could scarcely blame people because of his own inability to get to Cookstone before ten o'clock in the morning. The locals could hardly be expected to leave the dead man lying by the roadside all night. None the less, his number one rule had been broken. He liked to be able to stand where it had happened, to put himself in the murderer's shoes, feel the atmosphere. Staring at a stretch of flattened grass was not the same thing.

Well, he was here now, and people were waiting to see him. Perhaps they could add to his knowledge.

There wasn't much to be learned from the two flimsy reports and the handful of photographs which had been waiting on the desk. Walking across to the door, he opened it, barking,

"Right, let's get on with it."

Detective Sergeant Joe Grayson appeared, tapping at the open door more out of formality than necessity. Then he came in and stood before the desk, where the superintendent had seated himself.

Brent looked mildly surprised.

"Morning, Curly. Why you? Where's the inspector?"

Grayson cleared his throat.

"He's in court, sir. Four cases, I'm afraid. Shan't see him until the middle of the afternoon, at best."

Bloody fine start.

"I see. What do you know about this case?"

The standing man pulled a black notebook from his pocket and opened it.

"I was asleep at home when the call came, sir – "

"Time? – "

"Three-ten a.m. I went straight to Quarry Lane, where the body had been found. The ambulance was there, but there was nothing they could do, of course. The M.E. arrived a few minutes after me –"

"Few minutes? What do you mean, 'a few minutes'? What time did the doctor get there?"

Grayson turned over a leaf.

"At three-forty-two, sir."

"Took his time, didn't he? Why the delay?"

"He's not a police doctor, sir. Just one of the local G.Ps taking his turn on the rota. Joe Grayson had had dealings with Brent in the past. He knew from experience that he wasn't called Barker Brent for nothing.

"So you were at the murder site," prompted Brent. "All right for some. What did you make of it?"

Grayson swallowed. He'd known before he entered the room that there would be a fuss about the body having been removed without the superintendant seeing it.

"It was the body of a man about forty years old. Height five feet eleven – "

"I've read the description," was the curt interruption. "I asked you what you made of it."

"At the time, sir, not much at all. There was a certain amount of moonlight, but not enough for a proper look round. However," he continued hastily, sensing another interruption, "I went back out there at first light."

"Ah."

Brent pressed his fingertips together and stared up at his junior. Good man, this Grayson. It was lucky that he'd been the one on call-out. The superintendent could think of other names who wouldn't have suited his book at all.

"Right, so you went back this morning. I imagine we left a P.C. on guard there?"

"Yes, sir, we did. Well, as I say, I went back for a look in the daylight."

"And?"

"There was blood at the edge of the verge. Just a few spots. Then nothing for five feet two inches, when there was a smeared patch. This stretched for another seven feet, up until the place where the body was found."

Brent nodded.

"And what did you deduce from that?"

Grayson crossed his fingers mentally.

"It seemed to me that the man was standing at the side of the road when he was shot. He staggered forward slightly, then fell. After that, he either crawled a few feet, or he was dragged. I made a rough sketch of what could have happened."

Unfolding a sheet of paper, he held it across the desk. The superintendent took it, staring intently.

"It's possible," he muttered. "Yes, it's quite possible. If you're right, then the murderer would have had to be standing on the grass as well. Any traces there?"

"No, sir. As you know, there's been no rain worth mentioning for over a fortnight. The ground is hard. There's no question of getting casts of his footprints, I'm afraid."

"On the other hand," mused Brent, "just because our man was facing that way when he died, it doesn't follow that he was at the same angle when he was shot. In other words, he could have been facing any direction when the bullets went in. Either he could have turned, trying to run away, or the force of the bullets spun him round. In which case, the murderer might not have been on the grass at any time. What about the road surface?''

"Newly tarmaced, sir. No tyre marks, nothing that will help us.''

Brent lowered his head, staring upwards at the sergeant.

"Do you think there was a car involved?''

Grayson opened his hands expressively.

"If there wasn't, I don't see how the man got there. He wasn't exactly the walking type, and in any case he wasn't dressed for it. It's an isolated stretch of road, very few houses there. It's unlikely he came from one of those – ''

"Why?''

"We know all the people, sir. They've lived here for years.

He stopped as he saw the puzzled expression on his superior's face.

"What are you suggesting, Sergeant Grayson? That anybody who happens to have lived in Cookstone for a few years is incapable of murder? Am I to believe my ears?''

The younger man was discomfited.

"No, sir, I didn't mean that, of course. I've prepared a list of all the householders, with a few notes about each one. It's being typed for you now.''

"I shall read it very carefully,'' was the ominous reply. "All right, so what have we got? One well-dressed man, who got himself shot at some time between midnight and two a.m., according to this report here. The point is, what was he doing there, in such an isolated spot? He didn't get out of the car – all right, I'm assuming there was a car – but he didn't get out of it to admire the view. The view isn't much in the daylight, leave alone midnight. The other thing is, why is there nothing in his pockets to tell us who he was?"

"Presumably the murderer went through his pockets, removed anything that would tell us who he was so as to delay identification. Sir."

"Perhaps, perhaps. It won't do him much good, though. Unless this chap turns out to be one of those geezers who came straight off a merchant ship, after years overseas, like they do in the detective stories, it won't take us long to identify him. I don't get the murderer's point. Still, after we lock him up, we can ask him, can't we?"

It was one of the famous Brent jokes. Grayson contrived a smile.

"We'll certainly do that, sir. By the way, your mentioning the merchant ship reminds me about his hands."

"What about 'em?"

"They're soft. Very clean. Our man hasn't done anything like rough work for many years, if ever."

"H'm."

Brent picked up a photograph of the dead man's head and shoulders.

"Who are you, chummy?" It was as if he was

talking to himself. Then, "All this stuff has gone to C.R.O., I imagine?"

"Oh yes, sir. Went off this morning."

"Good."

One thing was certain. If the unknown man had any kind of criminal history, the Criminal Records Office would identify him in short order. Brent stood up.

"Well, let's go and take a look at him. Never did like photographs. They don't give you the feel of a man. Where've you got him?"

"At the Cottage Hospital, sir. It's a bit rustic, I'm afraid. They're not really equipped for – "

"They've got all the equipment he needs," interrupted the superintendent. "All that's required is somewhere for the poor sod to lie flat. Let's get down there."

At the hospital, a prim starchy woman was waiting at the entrance. Brent looked at the set of her features, and put on his most gracious smile. Grayson made the introductions.

"This is Mrs Cooper, sir, the administrator of the hospital. Superintendant Brent."

"How d'you do, Mrs Cooper." Brent stuck out his hand. "I'm afraid we've put you to a lot of trouble. We'll do our best to inconvenience you as little as possible. Eh, sergeant?"

"Absolutely, sir."

Some of the stiffness went out of her face.

"Good morning, superintendent. Well, it has been a little difficult – "

"Of course it has," he soothed. "Quite understand your position. Still, we're very fortunate to have you. As you know, a lot of cottage hospitals have been closed down. Far too many, in my opinion. Backbone of the Health Service, if you ask me. Take this case, for instance. We'd have had to cart the poor chap for miles if you hadn't been here. Would you be kind enough to lead the way, Mrs Cooper?"

What a nice man, thought Mrs Cooper.

What a flannelling sod, thought Sergeant Grayson.

There was an outside storeroom which was pressed into occasional service as an emergency mortuary. Mrs Cooper took them along there, and went about her duties. The corpse rested on a rubber-wheeled trolley, and was covered with a white sheet. Brent pulled the sheet carefully back and stared at the waxen figure. A thoughtful expression came onto his face, and he looked across at Grayson.

"Does he strike you as anyone we know?"

The detective sergeant took a pace forward and looked dutifully down shaking his head.

"Don't think so, sir," he denied. "Not one of our local customers, that's certain."

"No, no, I'm sure he isn't. Not what I meant. Further afield, something like the telly, or the newspapers perhaps. I've seen him somewhere. Anyway, you can't place him?"

"Sorry, sir."

Without replacing the sheet, Brent turned away.

"His clothes are handy somewhere, are they?"

"In this cupboard, sir. Except the valuables. I had those taken to the station."

The superintendent grunted something as he spread the dead man's clothes out on a trestle-table.

"Nice suit," he muttered. "not bad at all. Hundred and fifty quid, would you say? Two hundred? Oh, and the shirt's nice. He didn't get that in a supermarket."
Then he looked across sharply. "Valuables, did you say?"

"Nothing much, sir," demurred Grayson. "Wrist watch, gold fountain pen, some cash."

"How much cash?" was the immediate response.

"Thirty-four pounds, and some change. Hardly enough to commit murder for."

"Especially since the killer didn't bother to take it," pointed out Brent sourly. "What about car keys?"

"None that we found."

Brent began to stuff the clothes back into the white polythene bag.

"Let's assume for the moment that he didn't come by his own car. I mean, to this vicinity. If not, he had to come by public transport, and that won't take long for me to find out. Or if neither of those, then somebody drove him here from wherever he started out. We would have to assume, I think, that the person who drove him was the killer. Doesn't get us very far forward, does it?"

He stood, looking down again at the still figure, his face puzzled.

"On the other hand, sir," offered Grayson, "if he

did drive himself, then somebody might have an extra car stuck in his drive, or his garage.''

Brent looked doubtful.

"He's had all night to get rid of it, you know. Still, stranger things have happened. Make our job easy, eh?''

"It would, sir. The patrol cars are having a special look round this morning.''

"Good. Well, there's no point in standing around here. This chap's not going to tell us anything. I want to see the doctor next. He ought to be up by now. Have you got his home address?''

Grayson smiled faintly.

"He won't be at home, sir. He'll be doing his morning surgery.''

"I thought you said he was on the roster last night?''

"Yes, he was. But that is just an extra chore, lumped on top of his other work. Whatever sleep he loses, that just goes down to experience.''

Brent grinned in return.

"We're not the only ones, then. All right, get on the blower and find out when he'll be free. What sort of chap is he?''

Grayson thought carefully. He didn't like Keith Firbank very much. Too good looking for a start, and too damned sure of himself. None of the women in the village would agree with him, he knew, and perhaps that had some influence on his own opinion.

Brent noted the pause, and wondered about the reasons.

"He's quite young, sir, about thirty I would guess. Came here about two years ago, bought a partnership in a local practice. The patients seem to like him well enough. Especially the women."

The last words were out almost as he thought them. He hadn't intended to speak his mind on that score. The significance was not lost on his listener. To him, it was an old familiar pattern. The new young G.P., arriving into a close community, and the fluttering of a few hearts. It usually upset some of the males, and obviously Detective Sergeant Grayson was to be included among them.

"Reliable, is here?"

Brent was interested only in the medical man's capabilities, not in his impact on the lonely hearts column. This time, there was no hesitation in the reply.

"Most reliable, sir. Been very helpful to us, more than once. Did quite a bit of police work before he came here. Used to be on the staff of one of the big teaching hospitals in London. St. Francis, I think it was."

It was a good start. Brent had a jaundiced view of country G.Ps, with one eye on the stethoscope and the other on the golf-bag.

"Right. Well, whistle him up, will you? Tell him I'm sorry to interrupt his day, and I'll be as brief as possible. No chance of a word with the ambulance men, I suppose? They wouldn't have come from round here.

"Afraid not, sir. They were the emergency people from Wenford General. Would you like them brought over?"

Brent shook his head.

"Doubt if they could add anything. It would have been handy if they'd been on the spot, but it's not worth making a fuss over. What about the people on duty here last night? They'll all be home in bed, I expect."

Grayson swallowed. The superintendent wasn't going to be very impressed with the reply to that one.

"Only one person involved, sir. The night orderly. He went off duty at eight this morning."

"One?" Brent paused. "Are you seriously telling me that this miserable excuse for a hospital is not staffed during the night?"

"As I understand it, sir, there's no need. The patients here are not critically ill, mostly routine cases of care and rest. There are people on stand-by at home if an emergency crops up."

"Gawd help us," breathed his superior. "No wonder they close these places down if you're only allowed to be seriously ill in office hours."

He was being unfair, and he knew it. Sergeant Grayson knew it too, but was careful not to let his features betray his thoughts.

"I took a statement from the orderly, sir, before he

went off duty. It should be typed out ready for you by the time we get back."

They were interrupted by the entrance of Mrs Cooper.

"Sorry to intrude, superintendent, but you're wanted on the telephone."

"Ah, thank you. We're about finished here anyway."

She led him along to her small office and pointed to the waiting receiver. Brent picked it up.

"Superintendent Brent."

Then he listened, and his face changed. After a while he said, "There's no possibility of any mistake? Right. Thanks very much."

He put down the telephone, and nodded to Mrs Cooper.

"Thank you very much for all your help. I don't think we shall be troubling you very much further. Sergeant."

Mrs Cooper was disappointed not to learn what the telephone call had been about, but she smiled as the two men went out. When they were safely out of earshot Brent muttered,

"Well, Sergeant Grayson, our little murder might not be so little after all. They've identified our friend. Thought I knew his face, but I just couldn't bring him back to mind."

And you're not very pleased with yourself, thought Grayson. Out loud, he said, "Who is he then, sir?"

"A villain, sergeant. A real dyed-in-the-wool, black-hearted bloody villain. Name of George Parks. Does it mean anything?''

Grayson wrinkled his brow.

"Not right off, sir," he admitted.

"Try this, then. He's one of Blackie Webber's little friends. Don't tell me you've never heard of dear old Blackie?"

It was almost impossible to pick up a Sunday newspaper without reading about the alleged – always alleged – exploits of Webber and his crew. Grayson whistled.

"We are going to have some fun then, sir."

Brent nodded.

"That's one way of putting it. A top London villain comes to a sleepy place like this to get himself killed. Why, sergeant? Why?"

They were walking noticeably faster as they left.

Two

The door-chime was a special feature of the luxurious flat in Kensington Square. What was it the salesman had said— 'a whispered invitation to some mystic Eastern ritual'? Something like that. Certainly the sound was pleasant enough, a light but resonant gong. Still a bleeding doorbell, whatever you called it.

Blackie Webber scratched at the thick dark mat which was his chest, and frowned at his watch. Ten past eleven. Middle of the night. Somebody better have a good excuse, waking him up at this ungodly hour. He inserted a big elbow between the golden naked shoulder-blades beside him and pushed.

"You, Julie. Milkman's here. Go and answer the door."

There was a squeal of protest, and a strikingly pretty girl looked round at him, blonde hair falling across her face.

"Do what?"

"Somebody on the knocker. Go and see who it is."

"I haven't got no clothes on."

Blackie sighed. Birds were all right in their place,

but he couldn't stand them when they argued.

"There's a dressing-gown behind the door. Wrap that round you."

The girl was about to offer an argument, then saw the glitter in the hooded eyes. Swinging her legs out of bed, she stood erect, stretched and walked to the door. Blackie watched her movements, and the sleep left him.

"Tell you what, Julie, I'll get rid of 'em as quick as I can. Then you get back in here as quick as you can."

She tossed her head in disdain, reaching for the striped silk dressing-gown.

"We'll see. And it's Jeannie, not Julie."

"Yeah, well just get on with it, darling."

She went out of the bedroom, and Blackie got out of bed, making for the small refrigerator which formed one side of the ornate dressing-table. Removing a carton of milk, he poured some into a crystal tumbler, drank, belched mightily, and set the glass down. The day never got off to any kind of a start until he'd had his morning milk. Now, where did he leave those fags?

There was some kind of argument going on outside, a man's deep voice cutting across the girl's protests.

"'Ere, you can't – "

"Oh, yes I can, dolly. Well, well, here we are then, Blackie. And the top of the morning to you."

A man strode cheerfully into the bedroom, taking in the whole scene, and particularly the naked man by the dressing-table.

"You got a bloody fine cheek, bursting in 'ere. I'll

have the law on you."

The newcomer laughed.

"But I am the law, me old son. You remember me, Blackie. Chief Inspector Hudson, Serious Crime Squad."

He remembered him, all right. Nasty piece of work. Not like some coppers. They had their work to do, and that's fair enough, but you could at least talk to 'em. The Serious Crime lot, and this Hudson especially, they were nasty men. Had special powers, too, which made it worse.

"You've got no right, coming in here like this."

Blackie felt a bit of a fool, to tell the truth, rummaging around for his trousers. A man can't argue with anybody properly, least of all a copper, when he's dangling his posterities.

" 'E pushed me out of the way, Blackie," the blonde complained.

"Yeah, well, never mind, darling. Go and make a nice cup of tea."

He felt better now, buttoning the yellow silk shirt as he spoke.

"Tea?" The visitor did not sound impressed. "I can't drink tea in the middle of the day. Saw some nice brandy on the cabinet outside. Remy Martin, it was. I'll have a drop of that, thanks."

The girl looked at Blackie, who nodded, then she went out.

"Thought you weren't supposed to drink on duty," he accused.

Hudson smiled cheerfully.

"That's only for ordinary coppers. People who only work twelve hours a day. I work the whole twenty-four, and I'm not going teetotal, not even for the dear old Yard. That's a nice clock over there. French, isn't it? Where'd you nick it?"

Webber sighed, grateful for the patience he had learned to acquire over the years.

"I bought that clock proper legal. My solicitor has the receipt."

"I'll bet he has," was the retort. "Oh, are we leaving?"

His unsmiling host was walking out of the room.

"I'm not spending all day in the bedroom," he snapped.

The chief inspector followed him, grinning slightly. Sam Hudson was enjoying himself, and looking forward to the rest of the interview.

In the spacious living-area, Blackie settled himself heavily into a deep-padded leather chair, and waited. The policeman sprawled into its twin, which was placed opposite, and the two men sized each other up. They made an interesting contrast.

Hudson was over six feet tall, with shining fair hair styled fashionably, and not quite long enough to draw disapproval from his superiors. The face was round and cheerful, with a sardonic set to the mouth which prohibited any description of the baby-face variety. Always elegantly, though casually dressed, he was the very antithesis of what people expected to find in a man of his position. The general impression he conveyed was that of a fairly successful young actor, and that was the soubriquet awarded to him when he was not present. In his early days in the force, he had been found unsuited to almost every type of job he was given, and his superiors had almost despaired of him. His academic record was excellent, his keenness beyond reproach, but it seemed he was not to achieve the results looked for when he was first recruited. Finally, someone had suggested he be tried on the Serious Crime Squad, and it was quickly evident he had found his niche at last. After that, promotion followed rapidly, and at the ripe old age of twenty-seven he was already a chief inspector, and that was by no means to be the summit of his achievements. Sharp-witted, courageous, seemingly tireless, he roamed the streets and alleyways of the metropolis, to the eternal discomfort of the criminal element. All in all, he was, as the expectant Webber was thinking at that very moment, a very dangerous copper.

Blackie Webber was another proposition altogether. Five feet six inches tall, and broad as a barrel, he had embarked with gusto on a criminal career, trade-marked heavily by violence. Quick-tempered in his early years, he had often found himself in trouble on that score alone, not uncommon with people of his ilk. But, unlike most of them, Blackie had learned. There was no point in spending your life proving how tough you were, unless there was profit at the other end. He'd been fortunate, during a spell in Wormwood Scrubs, to share a cell with a highly respected criminal, twenty years his senior, who was always referred to as The Professor. Many a confinement hour had been spent in conversation with this paragon, who was serving his first jail sentence after thirty years of criminal activity, and even that had been due to betrayal. Blackie was fascinated to know how the Professor had managed to evade the law for all that time, and was therefore willing to listen to the older man's advice. It was simple, so simple that most people would ignore it, as Blackie himself would have done, under other conditions. But lying in that dark cell, planning the future, he had come gradually to recognise the truth of what he was being told.

"Weigh it up, son. There's two things to think about. First off, is there going to be any readies for Blackie out of it? That's the first thing. The second thing is, have you got a fair chance of getting away with it? Unless you've satisfied yourself about those two things my advice is this. Keep your hands in your pockets. And your sticker if you use one, or a shooter. Especially the shooter. You got to weigh it up, see?"

So spoke the Professor, and Blackie listened. He learned the lesson well. A few weeks before his release there was a brawl in the recreation room. The Professor was innocently involved and looked desperately round for his cell-mate to come to his aid. But Blackie was weighing it up. There was no profit in it for him, nothing but the prospect of having his precious freedom delayed.

The Professor was knifed in the mêlée, and badly wounded. Blackie Webber could have saved him, but he made himself scarce. He had come a long way since those days, combining a natural astuteness with his reputation as a ferocious adversary when he unleashed himself. He was adaptable, too. Prostitution, drugs, pornography, there was no area into which he had not moved at some time or other. He always moved out when he got bored, or when he sensed the danger signals. It had been many years since the police had been able to make out a case against Webber, although he was never far from their thoughts.

These, then, were the two men now seated on either side of a thick Turkish rug, while the pretty blonde trotted about with her offerings.

"Two sugars?" Blackie peered suspiciously at the cup she presented. "I always have two sugars."

"I know that, silly boy," pealed Jeannie, looking across nervously at the good-looking stranger. "Did you want anything in the brandy? Lime or anything?"

Hudson winced, shaking his head.

"No, thank you. Just as it comes."

She poured out a drink and took it across to him. He looked her over, trying to assess what shape she really was under all those voluminous folds. Top class, probably. Blackie Webber could afford the best.

"Well, this is nice," he said cheerfully. "A very cosy domestic scene. But you really ought to buy the missus her own dressing-gown, Blackie."

Blackie sipped noisily at his tea, while the girl

watched anxiously. To her great relief he nodded his satisfaction before replying to the jibe.

"If you've come here to represent the battered wives lot, you're wasting your time. Joannie is just a friend. Eh, darling?"

He smiled at the girl, who giggled squeakily.

"It's Jeannie," she corrected. "He knows, really. He only does that to annoy me. Pleased to meet you."

Hudson inclined his head pleasantly.

"You ought to be more careful choosing your friends, Jeannie. I could tell you terrible things about Mr Webber."

Blackie banged down his teacup decisively. The girl jumped.

"She hasn't got time now. She was just leaving when you rang the door."

"But, Blackie – " she began.

"You heard me. Piss off."

There was no further argument. Jeannie tossed her head and disappeared into the bedroom.

"And shut the door," was the barked order.

The door slammed, and Blackie smiled at his visitor without warmth.

"You didn't come here just for a free drink, I s'pose?"

Hudson shook his head.

"No, no. Just enquiring about old friends. How's Georgie Parks, these days?"

The face opposite became a mask.

"Parks? Georgie Parks? Do I know him?"

"You ought to. He runs out of your stable. Try to concentrate, Blackie. Only last week you spent two days with him at Kempton Park races. Three months ago you were both questioned about that bullion job in East Ham. Before that – "

"Ah," nodded Blackie. "That Georgie Parks. Yes, I've got him now."

"Bloody think so," snapped Hudson. "He's been on your pay-roll the past two years, that I can swear to."

"I don't have any pay-roll," denied Webber, unruffled. "Anyway, I remember this Parks, now. What about him?"

"When was the last time you saw him?"

The dark brows furrowed in simulated concentration.

"Let me see. Kempton Park last week, you reckoned?"

"Don't waste too much time, Blackie. I might get cross with you. Parks was with you yesterday afternoon, as I can prove from a dozen sources. What time did you last see him?"

While keeping up the delaying banter, Webber's mind was working furiously. Something seemed to have gone wrong, somewhere. If this clever bastard had any inkling of what was currently going on there could be a lot of trouble all round. Surely they hadn't nabbed Georgie? No, they couldn't have. Georgie hadn't actually done anything to do with the present scheme. Not yet. Something silly, now, that was another matter altogether. Georgie Parks could be a bit hasty sometimes, and he couldn't always control his temper. Perhaps he'd belted somebody and got himself in bad with the law. Whatever he'd done, it was nothing to the trouble he was going to have with one Blackie Webber next time he turned up.

"I'm waiting," reminded Hudson.

"Yeah. I know. Just trying to think," was the rejoinder. "Tell you the Gawd's truth, inspector – "

"Chief inspector."

"Right, yeah. Tell you the truth, I had a few jars last night. Memory's a bit whatsit, vague."

The policeman's eyes narrowed.

"A few jars? You? Never known you do that before."

"Somebody fixed 'em for me. I'll sort out whoever it was, I promise you. I'm remembering now though. It would have been about teatime. Yes. About that. Say five o'clock."

The bedroom door opened, and Jeannie came into the room. She looked incongruous in the sunlight with a near-topless red silk evening gown reaching to the floor.

"How can I walk down the street like this?" she wailed.

"Call a cab," instructed Blackie tersely.

"But I haven't got no money," she protested.

"Yes, you have, darling. Look in your purse."

The protest went from her face as she snapped open the black crocodile handbag and peered inside.

"Cor," she exclaimed. "Ta, Blackie, ta ever so."

"That thing over there – " a thick thumb jabbed in

the indicated direction – "it's called a door. When you go through it, shut it behind you."

She nodded, smiling tremulously.

"Well, ta-ta then. Goodbye, Mr – er – "

The two men stared at her impassively. Quickly she made for the door, and left them alone.

"Nice little thing," observed Hudson. "Regular friend, is she?"

Webber welcomed the change of subject.

"Met her at the church social," he explained. "We was going to go to choir practice this morning, but you've spoiled that."

"Really?" returned Hudson blandly. "I must have a word with your vicar. Funny sort of parish he's running. Now then, about Georgie Parks?"

"Parks? Oh yes, that's right. You was asking me when I last seen him. I told you. Yesterday. About five o'clock.

"So you did. And what did you do after that? Was that when you went to this church social?"

A man who has a lot of experience of questioning by police officers develops a sense of knowing when the time has come to watch his words. Blackie Webber knew that the time had arrived.

"What do you want to know for?"

"I'm nosey. Can't keep out of other people's business. It's a weakness I have. Well?"

Webber leaned back against the padded leather, relaxing himself. This Hudson didn't know anything. Not really. He could suspect what the hell he liked, but he didn't really know anything.

"Just tell me something. What right have you got to come in here asking me questions?"

Hudson sighed, recognising a time-worn ploy.

"Don't waste my time, Blackie. Don't make me go into some silly nonsense about warrants and all that jazz. You know I haven't got one, and you know it doesn't matter a toss. All that old guff is for ordinary police work and ordinary villains. I'm no run-of-the-mill copper and you're not some petty burglar. You're a very big operator, Blackie, a man of position. A real deep-dyed nasty bastard. People like you and me don't have to fart about with pieces of paper. Now, don't make me turn nasty on you. Just answer the question."

Webber pondered for a moment. If anyone else had spoken to him like that – but he dismissed the thought. Nobody in London would dare to. Nobody but this cocky, self-satisfied bleeder from the Special Crime Squad. The point was, what did he mean by turning nasty? There was nothing they could use against him that he knew of. Unless Georgie – no. Out of the question.

"How do you mean then, turn nasty? What's the charge? Or what would it be, I should say?"

Hudson opened his hands expressively.

"How about poncing?"

Webber was affronted, then amused.

"Poncing? Me? This is Blackie Webber you're talking to. You'll get yourself laughed off the force."

Hudson's headshake was emphatic.

"I don't think so. I found you here this morning, with a known prostitute – "

"That girl is not on the game – " interjected Blackie swiftly.

"Oh, tut, tut. I can produce people who will swear she is, and you know I can. Then before she left she handed you the money she'd earned last night. I witnessed that personally."

"Lying sod."

"Watch your language now. Think of the papers, Blackie. Think of the disgrace. The great Blackie Webber, terror of the West End, and all that bull. What is it going to look like? You tell me I'll be laughed off the force. I think you're wrong. I think you're the one. I think you'll be laughed off the streets."

"H'm. Like that, is it?"

It was all chat, of course. Blackie knew he would never allow things to get that far out of hand. Still and all, this geezer would have to be answered.

"No need to be nasty," he resumed. "I don't mind telling you where I was. Why should I care? I was at the Green Table Club, losing me life savings. It's a proper club, and I'm a proper member, card and all. Oh, and there's probably best part of twenty people who can swear to that."

"I'm sure there are," said Hudson, absently. "What time did you leave? Roughly will do."

Webber shrugged.

"Two o'clock. Half-past. About then."

"That's a pity," Hudson replied. "You can probably prove that, no doubt?"

"Spect so, if I have to. Why, what does it matter?"

Hudson sipped maddeningly at his drink before answering. He hadn't come to the flat with any real hope of pinning the Parks murder on Webber. The killing was completely uncharacteristic of what would be expected, and apart from that, the locale would present too many complications. If Parks had died in London, the interview would have taken a very different course. Still, it was always a pleasure to be in a position to ruffle Webber's feathers.

"So you're able to prove where you were between midnight and one o'clock last night?" he persisted.

"I just told you."

"Then you weren't in Cookstone at that time?"

The dark eyebrows lifted in slow surprise.

"Cookstone? Where's that? I never even heard of it."

Hudson smiled infuriatingly, spinning it out to the last.

"Lot of people haven't," he agreed chattily. "It's only a small village. Still, they'll all know about it soon enough. When they read the papers."

Blackie sighed.

"All right then, I'll buy it. Suppose you tell me what happened in this Cookstone between midnight and one o'clock?"

"Oh, didn't you know? Georgie Parks got himself scragged."

Even Webber's calm left him.

"Eh? What're you saying?"

"Georgie. Your old mate. Somebody bumped him off, Blackie. Murdered him. Thought you'd like to know."

Blackie's mouth dropped open.

"Stuff a little duck," he whispered softly.

Three

Keith Firbank yawned and rubbed at his eyes, producing an all-too-familiar sensation as of someone scraping sandpaper across the eyeballs. Ten-past one. He was due to start his afternoon rounds at two p.m. and there was this police superintendent to be seen before that. The message had said the chap – what was his name, Brent, that was it – Brent would make himself available any time from one o'clock onwards. Damned decent of him. He'd probably sat down to a respectable steak and kidney pudding or something in the police canteen.

He picked up the telephone and waited.

"Yes, doctor?" A woman's voice.

"Who is that?" he queried. You never knew who'd be on duty.

"Mrs Parfitt."

"Oh, Mrs Parfitt, I know it's a bit early, but are my rounds files ready yet? I have to call at the police station, and if I could take them with me it would save me coming back."

"Just finishing them now. I'll bring them in as quickly as I can."

"Thank you."

Mrs Parfitt, he puzzled, replacing the receiver. Was that the pleasant-faced elderly woman with the bun, or the blonde with the legs? So many of them out there, what with the part-timers and everything. Not that it mattered, if only the rounds files were correct and complete for once. There'd been a fearful cock-up one day last week, and even then it was almost impossible to allocate the blame.

So far as lunch was concerned, he'd stop at the baker's on his way to the police station. They did packets of sandwiches to take away, and he could eat them in the car in between calls. They'd be cheese again, naturally, they always were. Still, it would be better than nothing.

The door opened and MrsParfitt came in with an armful of files. She was the blonde with the legs after all.

"There we are, doctor. All present, I think."

She flashed him a smile. Nice teeth.

"That was quick," he said gratefully. "Thanks very much."

"Oh, that's all right, doctor. I know how busy you are. If there's anything else I can do, anything at all? There's no one else out there at the moment, you see. Lunch time."

She was eyeing him archly, and he didn't think she'd be much trouble. If he'd been back at St Francis' he'd have asked her how she would feel about five minutes' overtime on the stretcher. But this was not a big London General. This was a village G.P. set-up, and a very different kettle of fish.

"Very good of you," he demurred, "but I think I have all I need for the moment."

She took her time about leaving the room, and the view was pleasant, even to a very tired man. Ah well. He scooped up the files, checking them quickly against his handwritten list, and picked up his bag. All set for the road.

There was a rear door for use only by the doctors, and this was to ensure they did not get snared by patients as they came and went from the surgery. Dr Firbank walked out to his three-year-old Cortina, and stacked the files in the boot. Then he drove down the High Street to where the police station occupied an unobtrusive position.

The desk sergeant knew him by sight.

"Dr Firbank isn't it?"

"That's right," he confirmed. "I think your Superintendent Brent is expecting me."

"I see, sir. Well, you'll find him using the governor's office. Second door to your right."

"Thank you."

He went down the passage and tapped at the second door, opening it at the same time. A dark-haird man not much older than himself looked up at the interruption.

"Superintendent Brent? I'm Dr Firbank."

"Ah."

The superintendent rose, holding out his hand.

"Good of you to come, doctor. I know how busy you are."

This man was no village bobby, decided Firbank, settling himself in the indicated chair. He would have been interested to know what at the same moment Brent had decided that he was no country practitioner.

"Now then, doctor, before we start, you've come straight from the surgery I believe?"

"Yes, as a matter of fact. Just issued the last half-gallon of the pink medicine."

Brent smiled.

"Three times a day after meals?"

Firbank smiled in his turn. Pleasant chap, this.

"You won't have had any lunch, then. Neither have I. No more time on my job than you have on yours. Hope you don't mind, I took a chance and laid something on for both of us while we're talking. Nothing fancy, you understand."

"Mind? Lord, no. I shall be most grateful."

It wouldn't have to be very fancy to be an improvement on those interminable cheese sandwiches. The superintendent must have pressed a buzzer under the desk or something, because the door opened almost at once and a woman constable came in with a tray. Beef sandwiches, sausage rolls, and a couple of bottles of lager. Brent began to pour out the beer.

"Lager all right? It's all we can offer I'm afraid."

"Excellent. Thanks very much."

"Please help yourself to the food."

Brent set an example by picking up the nearest sandwich and biting into it.

"I understand you put in some time at St Francis' before you set up down here."

Firbank swallowed some of his sausage roll, nodding.

"Best part of five years altogether. Of course, the earlier years were mostly finishing off my training."

"M'm. Busy part of London, though. Traffic accidents and the like. Quite a bit of stuff in our line too, I've no doubt."

The doctor lifted his glass, said "Cheers," and sipped at it.

"Oh yes, I can claim a fair experience of that line of work. Muggings, pub fights, football crowd stuff. Never a dull moment in dear old Emergency."

The superintendent grinned, returned the salute, and drank some of his beer.

"Had much to do with gunshot wounds?" he queried.

"Not much. The boys up there seem to favour more hand to hand stuff. Knives, razors, broken bottles. Not many guns about, even today."

"But some?" pressed Brent.

"A few, yes."

Doctor Firbank was wondering what the conversation was leading up to.

"Has anybody brought you up to date on this murder of ours?"

"Up to date in what way?"

"Has anybody told you who the victim was ?"

Firbank shook his head. Brent looked satisfied.

"Ah. Well, a murder's always a bit of a mystery, but this one's more so than most. The dead man is, or was I should say, a well-known London villain. Name of Georgie Parks."

That raised the visitor's eyebrows.

"Really? Then what was he doing down here? Not exactly the West End, is it?"

"Precisely, doctor, precisely. What indeed was he doing down here? Do you know, I'm almost as intrigued about that as I am about the murder. Nothing down here to interest a chap like the unlamented Mr Parks. Anyway, that's my worry, and I'll find out in the end. What I want to ask you, if it's not an unfair question, is whether you think those bullets would have killed him outright. I mean, would he have lived a few seconds, a few minutes, as long as half an hour, or what?"

"M'm."

Firbank chewed slowly on his sandwich, and thought about it.

"You appreciate I didn't conduct a proper post-mortem on the body? I mean, my job was to ensure there was no life left in the man, and certify him as dead. The coroner will produce a much more detailed and reliable report."

"I appreciate that, doctor, but I'm a great believer in immediacy. The gut reaction of the man on the spot. I've found it to be an invaluable pointer many times in the past. Let me tell you what I'm wondering. Georgie Parks had three bullets put into him, all in the chest and stomach areas. The point is, how long would it have taken for him to die? Would it have been possible for him to be shot somewhere else, miles away perhaps, then driven here by car and dumped at the side of the road where our man found him?"

Firbank frowned, thinking about this new reasoning.

"A lot would depend on the size of the bullets," he muttered, almost to himself.

"They were thirty-eights," he was advised crisply. "A fairly heavy slug for a man to take."

The doctor smiled apologetically.

"I'm sorry, but that doesn't mean a great deal to me. Ballistics is a subject on its own. The only thing I know about handguns is that air pistols use .22 ammunition, and the cowboys have forty-fives. Outside of that I'm fairly ignorant. However," he went on, seeing that Brent was about to come back at him, "since you say they are heavy bullets, then I don't see how the dead man could have survived very long. As you say, his chest and abdomen were a fearful mess. Of course, there's another possibility – no, I'd better keep quiet. You've got me forgetting I'm talking to a senior police officer."

Brent shook his head.

"Don't let that worry you. I'd be interested to know what flashed through your mind then."

"Well," began Firbank apologetically, "I was thinking of a compromise between the two alternatives. You asked whether he could have been shot somewhere else and driven here afterwards, as distinct from being shot where he was found. Is it at all possible that he was shot while the car was actually in motion and then just pushed out of the door to die? It would mean someone has an awful mess in his car, of course."

He looked across at the listening man.

"Like in the old gangster movies, you mean?" mused Brent. "Possible, of course, but it would involve two men instead of one. You'd have to have a driver, as well as the one who fired the gun. Still, it's a thought."

The superintendent seemed to have reached the end of his questioning, and that suited Keith Firbank very well. After all, he wished to be as helpful as he could, but the kind of information which would really assist the police could only be produced by the in-depth forensic work performed by the coroner. He wondered fleetingly whether he should mention the other business, about the night orderly. Didn't want to get the chap in trouble, and that made him reluctant. On the other hand, this was a case of murder, and perhaps he would be doing the wrong thing by keeping quiet. The chap had certainly seemed to him to be looking rather guilty. He made up his mind as he swallowed the last of his beer. He would slide in at an angle, as his rugby coach used to advise.

"You were very quick with the identification, superintendent. I understood from Sergeant Grayson that there was nothing on the man that would help you."

Brent smiled broadly.

"No driving-licence, you mean, things like that? No, there wasn't. But whoever killed Parks was wasting his time with stuff like that. He was still wearing his dabs. So far as we are concerned, that's as good as carrying a sworn statement of identity, signed by three J.Ps and a vicar. We knew who he was in a matter of hours."

"Dabs? Oh yes. Fingerprints, of course."

The superintendent inclined his head.

"There have been cases where people have done terrible things to bodies to get rid of those items. I leave the details to your imagination, but you can believe me, those methods are not for the squeamish. Our man either didn't fancy that kind of work, or he just didn't have time."

Firbank nodded.

"Yes, nasty business, I can imagine. Perhaps the murderer wasn't really so concerned about removing identity papers. It could have been just accidental that those things were all in the dead man's wallet. When he removed the wallet, he removed the driving-licence or whatever without particularly intending to."

Michael Brent stared thoughtfully at the last remaining piece of his sausage roll, then popped it into his mouth. There was something in this, he could feel it. The doctor wasn't just making conversation. He was fishing, and Brent could always sense when someone was trying that on. Perhaps there was something here for a simple copper to learn, and Brent was always in the market for fresh knowledge.

"What makes you think he was robbed, doctor?" he asked mildly.

The man looked discomfited for a brief moment.

"Oh, nothing really," he denied airliy. "I suppose I just assumed that a murderer would take whatever he could find."

He was definitely being evasive now. Brent wiped his fingers happily on a spotless handkerchief. He was about to learn something.

"By no means does that follow," he assured his listener. "And there was no question of it here. All valuables intact. There was even thirty odd pounds in cash."

"Really? Oh, well that knocks my idea on the head then."

The words were innocuous, but why should there be a certain amount of relief in the doctor's tone?

"That obviously surprises you," Brent said sharply. "Why?"

Firbank didn't exactly blush, but he did look faintly guilty.

"No, not really," he denied. "Well, to tell you the truth, I'm a bit relieved."

Now we'd have it.

"Mind telling me? Relieved about what?"

The doctor swallowed, then made up his mind.

"That there's nothing missing," he admitted. "You see, when I first go to the hospital, Albert Spence was –"

"Spence? Who's he?"

"The night orderly," was the explanation. "Well, he was standing near the dead man's clothes. He'd undressed him, you see. Had a lot of experience removing clothing from corpses. It's not as easy as it might sound."

"I understand. Please go on."

Firbank shrugged.

"Well, it was obviously nothing. Just my imagination working at half-past three in the morning, or whatever it was. It seemed to me that he looked guilty somehow. Well, perhaps not guilty, that would be too strong. Ill at ease. Yes, ill at ease, that's a better description. I thought to myself at the time, I wonder whether our Albert's helped himself to this chap's cash, or his watch or something."

"Why should you think that? Has he got a history of that kind of thing?"

"No, he certainly has not. Very well thought of around here, and with reason. It does me no credit that I allowed myself such a thought, and I can tell you I'm highly relieved to be wrong."

Brent nodded pleasantly.

"Easily done, doctor. You're the medical man here, so I don't have to tell you what happens to the imagination after a long spell with no sleep. Very familiar to me, I assure you."

His visitor rose to go.

"Well, I'm glad it's off my chest. And now, superintendent, if you'll forgive me, I'll be on my way. For you, the world is full of crooks. In my case it's the sick and the palsied. Thanks for the lunch. Enjoyed that."

Brent rose, holding out his hand.

"We must do it again. Somewhere a bit less official."

The doctor left the office, and the superintendent picked up a telephone.

"Put Grayson on, please. Curly, is that you? Right. Take a car, and get out to wherever this Albert Spence lives. The night orderly chap. If he's in bed, get him out. If he's gone out, find him. I want him in this office within the hour."

Leaning back in his chair, he suddenly rubbed his hands together. Anyone with experience of working with him would have recognised that as a sign of his extreme pleasure.

The telephone rang.

"Superintendent Brent," he announced.

"Got a call for you, sir. From the Yard."

It was evident from the switchboard operator's tone of hushed reverence that calls from Scotland Yard were not an everyday occurrence at the Cookstone nick. Superintendent Brent, however, was not so readily impressed.

"Well, put 'em through, man."

There was some clicking, and a pleasant male voice asked, "Is this Superintendent Michael Brent?"

"Yes. Who is that, please?"

"My name is Hudson, Sam Hudson. Chief Inspector, Serious Crime Squad."

Chief Inspector, eh? Well, that wasn't so bad. At least he wouldn't have to click his heels. In fact, strictly speaking, any heel clicking ought to come from the other end, but he knew from experience there wouldn't be much chance of that. Not with those Serious Crime tearaways.

"What can I do for you, Chief Inspector?"

At the other end , Sam Hudson pulled a face. There wasn't going to be any Sam and Mike about this conversation, by the sound of it.

"I hear you've got one of my little playmates down there, Georgie Parks."

So that was it. Those bloody people got their noses into everything.

"That's right," confirmed Brent. "But I'm afraid he can't come out to play today."

"That's what I hear," returned the Yard man. "Who've you got locked up for it?"

"Bit early for that. Haven't even had the coroner's report yet." Then a thought came into Brent's head. "Matter of fact, I was thinking of giving you people a ring a bit later on."

That'll be the day, thought Hudson inwardly.

"Really? What about?"

"I wondered whether you might have any idea of what Parks has been up to lately. Anything that might help with the investigation."

"Nothing at the moment, I'm afraid. I've got the feelers out, though. Might come up with something. Went and saw Blackie Webber this morning. I wanted to see his face when he heard the good news."

"And?"

"And nothing. Well, nothing incriminating, that is. Our dear friend was bowled over, you might say."

"Huh," snorted the superintendent. "How can we be sure he wasn't involved?"

"Well, we can't be absolutely sure, naturally. But he certainly did not do the job himself. I've checked out where he claims to have been at the vital time, and there's no question about it. His alibi is watertight. Still, I'm keeping my ears to the ground. If anything comes up that looks as if it might be of any help to you, it'll be passed on, and fast."

"I'm much obliged to you for your interest, Chief Inspector. Glad of any help I can get on this one. Looks like being a right little mystery."

Sam Hudson selected his words with care before he spoke again. These bloody provincials could be so huffy if offended. And the possibility of any unsolicited Yard assisstance would give offence in record time.

"I wonder, while we're on, whether you'd mind telling me if there was any indication about Parks' recent activities? I don't mean in connection with what happened to him, necessarily. The fact is, you see, that Blackie Webber has been very quiet lately, and it's been worrying us. Normally, we know roughly what he's up to, and we keep an eye on things, looking for a chance to nail him. Recently, he seems to have been doing nothing at all, and that is most uncharacteristic. The Parks affair, of course, is your department. Goes without saying. But if you happen to pick up anything that might give us something to work on I'd very much appreciate it."

There. That ought to satisfy Mr Brent. The Yard didn't usually go out asking for help like that.

Hudson was partly right. Superintendent Brent was considerably mollified by the easy friendliness, and the unspoken assurance that there would not be any meddling from London. If he could help them out he would be glad to do it, would have done it in any case, as a matter of course.

But he still had his reservations. This chap Hudson sounded all right, but he was still a Serious Crime merchant, and they had their own unorthodox little ways. In reply he said,

"I'll be pleased to do whatever I can, naturally. By the same token, Chief Inspector, if anything drops into your lap that you think might help a poor country copper, you'll pass it on?"

Poor country copper, forsooth. Hudson grinned at the telephone. He'd heard about this Brent. Bit of a dynamo, according to the reports.

"Automatically," he said cheerfully. "Perhaps we'll bump into each other during the course of this lot. Be nice to have a drink and a chat.''

"Let's hope so," returned Brent equably.

They said their goodbyes and hung up.

The man in Cookstone stared thoughfully at the now silent telephone. He was wondering how long it would be before their interests clashed.

The man in London was wondering the same thing. But that was for the future, something that could wait. For the moment, he had this more pressing worry on his mind. What he had told the superintendent was quite true. Blackie Webber had been too quiet for too long.

What the devil was he up to?

Four

What Blackie Webber was up to, at that particular moment, was lying in an enormous chocolate-coloured bath, and pondering on the unwelcome news he had received. There was no personal grief involved. Georgie Parks worked for him, and now Georgie Parks was dead. End of story. It was a pity though, because he wouldn't be easy to replace. The hairy man in the tub was in the same situation as any other executive in the world of business. He had lost a trusted lieutenant, and now he would have to face a recruitment problem. This was always a difficult task in the upper echelons, made even more complicated when your activities were mainly crooked. You didn't walk into any pub frequented by the fraternity and take on a man because you liked his face. There were more important considerations.

Georgie had been reliable. That was first and foremost. He had also been very intelligent, very knowledgeable about the game, and he mostly had control of his temper. Blackie had learned to apply the principle acquired from his old mentor, the Professor, to those people who came into his team. With some people, of course, the so-called control had a different basis. They just didn't have any bottle, but that was far from true in Georgie Parks' case. Georgie could be a very rough handful in a set-to, as he had demonstrated more than once.

But his main asset was his reliability. You could trust Georgie. If he had work to do he wouldn't be out getting drunk, or with some bird. Work was first, with him. And he wouldn't go around flapping his gums, another great weakness with people in this business.

No, it was going to be very hard to find another Georgie.

As to the present, there'd be one or two practical things to deal with. He'd have to satisfy himself that Georgie had not left anything lying around which might cause trouble. Someone would have to run the comb over that bird's place, the one he'd been shacking up with lately. It would take the law a little while to find out about her, so there was no mad rush. Still, it had to be seen to.

Then there was his widow. Blackie would have to send her a few quid, which was more than Georgie ever did. You had to show a bit of respect on these occasions. He'd go and see her personally, that would be a nice touch. Poor cow, she couldn't have had much of a life, and she could look forward to even less of a future. Well, no concern of his.

As to the question of Georgie's murder, that was no

concern of his, either. The police got paid for that sort of work. Let them get on and do it. Keep 'em busy, and perhaps give other people a bit more time to get on with their own affairs. In a way, if Georgie was going to get himself scragged, he couldn't have picked a more convenient time. There was nothing in Blackie Webber's present way of life that wouldn't bear the closest examination, even by the police.

For Mr Blackie Webber was taking a sabbatical.

"A what?"

He'd been in conversation with some geezer who seemed to be some kind of a schoolmaster, when he first heard the term.

The other man smiled.

"A sabbatical," he repeated. "I've been at the university now for twenty years, and a man can get stale. Needs freshening up. So I've taken a year away from the place, and I intend to put it to good use."

You're off to a flying start, mate, reflected his listener. They were at the bar of a particularly expensive West End haunt, and the stranger had struck up a conversation with him. Blackie had been amused when it started. This posh twit would have done his knickers if he'd known who he was talking to. But this whatsit – sabbatical? – intrigued him.

"What about money?" he enquired. "Not many people could afford a year off work, stale or not."

"Ah," smiled the stranger, "I would be one of those people, in the ordinary way of things. This is recognised as an occupational necessity. I shall continue to be supported."

Supported?

"You mean, the firm'll pay your wages even though you're not working?"

"Well, half-pay, actually. But with income tax the way it is, the net amount won't be very much less than I'm accustomed to."

Blackie had been pensive for days after that casual encounter. A year off, the man had said, on half-pay. It was a lovely idea. A nice long break. No scheming, no looking out for new angles, no worries. As to the money side, well, that would be no problem. He was a man who liked his comfort, and he lived well. On the other hand, he wasn't a fool like so many of 'em. Rushing out to Spain, or Tangier, anywhere it was hot, and drinking themselves stupid. Throwing big parties for hundreds of so-called friends, all that rubbish. Broke to the wide after a few weeks, and looking out for new opportunities to put their freedom in danger.

That wasn't Blackie's way. He put a certain percentage of his money to work for him, and now had a steady income from several sources, mainly legitimate. What with that, and the money he had in various strong boxes, he could take a long break without any financial hardships. Not a year, mark you. It was too long. A man could get used to not working and find himself doing nothing the rest of his life. That wouldn't do at all. No, six months was the limit.

Now, here he was, just a few weeks into it, and some bleeder had to upset everything by rubbing out old Georgie. It was really too bad. The police were a bloody nuisance at the best of times, sniffing about and annoying people. That was part of the game when you were working. What that university bloke would call an occupational hazard. But to have them sticking their big noses everywhere when you hadn't done anything at all, well, that was a liberty really. That Hudson, now. A nasty man, him. You'd think a good-looking bloke like that would take up some other line of work. They called him the actor, even his own mates. Why didn't he go on the telly or something? It was deceitful, that's what it was, him being a copper. And no ordinary copper at that. He'd put away some good people already, and he was only young yet.

Wouldn't it be lovely, thought Blackie fancifully, if Hudson cornered the geezer who'd done away with Georgie, and the geezer got cross and gave him some of the same. That was a nice thought. Now, if that happens, we'll go round with the plate for the bloke's defence. Be worth few quid to be shot of Mr Actor Sodding Hudson.

His reverie was interrupted by the ringing of the telephone which stood on a small ivory plinth beside the bath. Most of Blackie's ideas of comfort and luxury were lifted directly from the movies and sometimes the adverts on the telly. He'd always promised himself a telephone in the bathroom. There were two drawbacks he'd never considered. The first was that people would interrupt his bath. The second was that you got soap all over the bloody thing when you answered it. The people on the telly never did that.

"Well?" he demanded gruffly.

"Mr Webber?"

It was a woman's voice, strained and anxious. A cultured tone, though. Not the kind he was used to.

"This is Mr Webber," he agreed. "Who's that?"

"We met once. I'm Olivia Marshall. George Parks introduced us. Do you remember?"

Yes, he did. This was the posh bird Georgie was shacked up with. She didn't have no business phoning him at home. S'pose that bleeding Hudson was listening.

"What do you want, darling?" he said ungraciously.

"I'm worried sick, Mr Webber. They just said on the news that a man named George Parks had been murdered. Somewhere out of London, they said. I didn't catch the name of the place. The point is, George didn't come home last night. I wasn't too concerned, because I know he's – busy – sometimes. But I got scared when I heard the radio. It's not George, is it, not my George? Do you happen to know where he is? I wouldn't bother you if I wasn't so worried."

Great. This was all he needed, having to break the good news to some bird. Wait a minute, though. She might be useful. Putting on his most grave tone, he said,

"I'm sorry, Olivia. It's bad news, I'm afraid. It was Georgie who got done in last night."

There was an anguished wail at the other end, then deep sobbing.

Blackie held the receiver away from his ear, waiting for the racket to die down. Who would have thought it, eh? A bird like this, a quality bird, getting all worked up over old Georgie. He hadn't been much of a ladies' man, but him and this one had hit it off, and no mistake. Nice-looking sort too, he remembered, right out of Georgie's class altogether.

She was speaking again, jerking out the words in between strangled sobs.

"Who could have – why would anyone – Oh, Mr Webber, what am I going to do?"

Blackie stared at the creamy soapsuds lying on top of the black hairs on the back of his hand. Funny, the way the hairs kept the soap away from the skin, unless you rubbed it in.

Do? What was she going to do? Or, to be more exact, what was he going to do with her? It would be a little while before the law caught up with Olivia Marshall, but not all that long. Birds could cause no end of trouble for people, and this one would be worse than most, because she was a straight girl. The coppers would make mincemeat of her in no time. She'd have no idea how to behave. One hour with this little darling and they'd know the last time Georgie cut his toe-nails.

Blackie's instinct was to slip her a few hundred and

tell her to catch the first plane out of Heathrow, destination immaterial. Just keep her out of the country a few weeks till things settled down. But no, that wouldn't do. This girl wasn't just any old slag. This one would have a job, a family to worry about her. Bleeding nuisance, that's what she was.

"Mr Webber?" came the anxious voice.

"I'm still here, darling. I'm thinking. This has been a terrible shock, you see. Terrible. I'm trying to think what's best for everybody. Where are you now?"

"I'm at the flat. It's number sixty-one – "

"Yeah," he cut in quickly, visualising the dreaded Hudson with his pencil nicely poised. "Yeah, I know where it is. Tell you what. How would it be if I came round there, and we had a little chat. 'Bout half an' hour suit you?"

There was silence at first. Then she said tremulously, "I don't know. I mean – er – "

"Oh, I see."

The silly cow thought he was going round there for a free ride. Well, you couldn't blame her, really. That's what a lot of blokes would have done. Might give her a go himself when all this was out of the way. But not now. Business was always first.

"You've got the wrong end of the stick, Olivia. I don't mean you no harm. I want to help you, if I can. We've got to sort things out for Georgie."

The mention of the dear departed ought to do it, he reflected smugly

There was a fresh outburst of weeping at the other

end, then, "Yes. Yes, of course. I'm so confused. Please do come, Mr Webber. I'm not usually such a fool, but I'm really a bit lost."

" 'Course you are, dear, course you are. Now, go and make a nice cup of tea, and I'll be there as quick as I can."

Always give 'em something to do. It occupies their hands, and gets their minds off the problem. Blackie looked up at the hideous gold clock.

He hadn't got much time.

* * *

Albert Spence stood nervously in front of the desk, wondering when the superintendent was going to look up from the papers he was studying. Albert didn't know much about the police, but he knew this superintendent was a very big wig indeed, come all the way from the County Headquarters.

Sergeant Grayson stood beside the waiting man, unspeaking. That was odd, the way Mr Grayson had spoken to him when he came. He was always such a cheerful sort of bloke, easy to talk to. Not today, though. Very crisp, very formal. You'd have thought he was arresting Albert for the murder of that poor bloke.

Here, it couldn't be that they thought –

"You are Albert Wilfred Spence?"

The use of his full name brought Albert quickly back to the present. There was nothing friendly about the superintendent's face, or his voice.

"That's right, yes," he agreed.

"You are employed at the Cookstone Cottage Hospital in the capacity of orderly?"

"I am at the moment. Usually – "

"And were you on duty in that capacity last night?"

"Yes."

They all knew this. Made enough fuss about him making a statement and everything. What was this –

"And in the course of the night, were you called upon to assist with the reception of a deceased male person?"

" 'Course I was," he protested. "It's all down in the –"

"And in giving that assistance you had access to the belongings of the dead man?"

Albert became tight-lipped. It was no use trying to talk to this bloke. All you could do was say yes or no.

"Yes," he agreed stiffly.

The superintendent's eyes swivelled to Curly Grayson.

"Detective Sergeant Grayson, you are a witness to these admissions?"

"I am, sir."

Admissions? What were they on about? All this stuff was on the record. Blimey, he'd signed the thing in three places.

"What're you talking about, admissions?" he blurted out.

"Oh." The superintendent stared at him coldly. "What have we now? A denial? You wish to retract your statement?"

"Certainly not," stammered Albert. "I'm not admitting anything, I mean, I haven't done anything, I mean – "

His voice faltered away in his confusion.

The superintendent picked up a piece of paper with some sort of typed list on it. Albert thought it looked familiar.

"This is a list of the items certified as having been removed from the person or clothing of the dead man. P.C. Hibbert prepared this, did he not, Sergeant Grayson?"

"He did, sir," confirmed the sergeant.

"P.C. Hibbert's signature is at the foot," continued Brent, ponderously. "Along with a second signature, which looks to me like A.W. Spence. Is that correct?"

Albert nodded.

"Yes, me and Tom signed it, after we – "

"Tom?" The superintendent's face was awesome. "Who is Tom?"

"Tom Hibbert," replied Albert unhappily.

"I see. You and P.C. Hibbert are friends, then, are you?"

"I know him, yes."

"Trusts you, does he?"

"Well, I would hope so, after all these – "

"Trusts you enough to let you get on with what you were doing, without watching you all the time, eh?"

"Don't know what you mean."

But there was nervousness in Spence's voice now, and the new note was not lost on the superintendent. Brent nodded heavily, laid down the paper and picked up another sheet.

"I have another list here. A list of items known to have been in the possession of the deceased at the time of death. A very different list. The two lists do not match up at all. One thousand pounds in five-pound notes. One gold ring with an emerald surrounded by diamonds, approximate value fourteen hundred pounds, one silver chain; there are other things. But then, I expect you know the items as well as I do. Better, perhaps."

Money, diamonds, what was he on about? Albert swallowed hastily. This was turning out very nasty indeed.

"Look here, I don't know nothing about those things. What're you getting at? What're you suggesting, eh?"

Michael Brent widened his eyes in mock astonishment.

"I should have thought that was perfectly obvious. In a few moments I shall charge you formally with the theft of these items, after which you will be locked up."

"Locked up?" Albert couldn't believe his ears "You've got no right to charge me. I haven't done nothing."

"I know a lot about juries," mused Brent, as though talking to himself. "They're an unpredictable lot quite often. But there are some things they will not stand for. Robbing the dead is high on the list. You're in a lot of trouble, my friend. It was lucky for us you were seen. Otherwise we might have had trouble getting a conviction. As it is – "

He shrugged, to indicate that the whole business would be no more than a formality. Albert Spence had never been so frightened in his life. Witness? How could they have a witness to something he hadn't done? There was only that business with Dr Firbank. He wondered at the time whether the doctor had spotted – but that had nothing to do with diamonds and all that.

"I'm innocent," he protested shakily. "You can't just drag a man out of bed and accuse him of any old trumped-up charge."

The superintendent's smile was worse than his scowl.

"And you'll swear to that, naturally."

"Certainly," was the immediate reply.

The chief inquisitor considered for a moment, then, "Sergeant Grayson, have you known this man very long?"

"A few years, sir."

"If he gave you his word not to attempt to flee from justice while our enquiries were proceeding, would you be inclined to accept that?"

It was greatly to Grayson's credit that he contrived not to laugh at the image of Albert Wilfred Spence in this new rôle of desperate criminal and fugitive from justice. The hapless Albert waited anxiously to hear what he would say.

"Well, sir, I think yes. On balance, I would be inclined to."

"Subject to a sworn statement?"

The sergeant had no idea what his superior was up to, but he knew how to back him up.

"Always subject to that, sir, of course."

That seemed to satisfy Brent, who nodded, opening a drawer.

"Very well. I've had a statement prepared. All you have to do is sign it, with your hand on this Bible, and read this oath. Perfectly legal, I am an officer of the court."

Spence's face was perspiring with joyous relief. He took the proffered pen and leaned over, reading the typed words. Then some of his new happiness left him, and he hesitated.

"This doesn't say anything about money or diamonds and that," he objected.

"No," agreed Brent smoothly. "If we went into all that detail a good lawyer might talk you out of trouble, just because one piece was wrongly described, or something. This is just general wording, nobody can wriggle out of that. It simply states that you swear there are no items deficient from the list you and P.C.Hibbert signed this morning. In fact, that there was nothing else."

Spence bit his lip.

"I don't know nothing about no diamonds and that," he repeated doggedly.

"So you keep saying." Brent was impatient. "So all you have to do is sign that, make your oath, and you can be on your way."

Albert Spence's entire criminal history consisted o

one parking offence eight years earlier. Now, he could visualise the whole panoply of the law ranged against him. Sarcastic barristers in wigs, judges in red robes seated high above the floor. The jury, seated in silent disbelief.

"Albert Wilfred Spence, you have been found guilty by a jury of your peers – "

"You could make a crime out of anything if I sign this," he said weakly.

"But there isn't anything," insisted his tormentor. "You've said so often enough."

Albert laid down the pen.

"Oh no, I ain't signing nothing. You're not going to lock me up for one lousy little thing."

Brent lowered his head, in case triumph could be read in his eyes.

"What do you call one lousy little thing, Albert?" he asked softly.

"A match cover, that's all. Not even in perfect condition. Three matches gone out of it."

The superintendent looked up then.

"Match cover? You pinched his matches? What for, hadn't you got a light?"

Spence shook his head.

"Don't smoke," he denied. "It's not that. I collect 'em, you see. Never seen this one before, and he wasn't going to want it, was he? Not exactly the sort of thing the family would fight about either. No good to anybody. Don't see what the harm was. But I ain't signing this."

He seemed prepared for an argument about his statement, but Brent had lost interest in it.

"Unusual, was it? Describe it to me."

"Black and gold cover, name of some place in London on the outside. Can't bring it back for the minute."

"I want it on my desk as fast as a police car can get to your house and back. Go with him, sergeant."

"Sir."

Grayson turned to go, but Albert hung back.

"Just a minute, what about the oath and everything?"

"Albert, son, I'm going to give you a little bit of advice," breathed the superintendent. "At the moment, I'm only interested in this match cover. If you hang about here I'll have you on about ten charges, starting with obstructing the police in the course of their duties. Get out, while the going's good."

"Come on, Albert, while your luck holds," advised Grayson.

The little man scuttled out hurriedly.

Brent smiled to himself. This could be a breakthrough.

Five

The murder in Cookstone Village was very welcome copy for the midday newspapers, and they had all found room for it in early editions. Without very much in the way of hard information they had mostly done an excellent job of padding, with plenty of local interviews and faithful logging of every movement by a police car, involved or not. But the lack of identification of the dead man was a nuisance, although they did what they could with the 'mystery victim' angle. The fact that the murdered man was a well-known London villain was not established in time for the early editions, and some hasty reshuffling was necessary for the late extras. The following day, of course, the dailies would have a whale of a time, and many fruitful hours in which to set the story up, but that was always the hard luck of the midday and evening editors, as any newspaperman will tell you.

By the middle of that first afternoon, the most reliable news medium in the world had carried the message far and wide, that medium being the time-

honoured jungle drums. In pubs and clubs throughout
the metropolis the word had gone out about the fate of
Georgie Parks. Speculation was rife as to the identity
of the killer. There was other talk, too, about the gap
in Blackie Webber's team, and whether this might
weaken Webber's status. Some of the younger element
thought it might, with the exuberance of youth. The
older and more experienced heads took a more
thoughtful stance. Blackie was not the man to be
toppled because of one blow at his empire, serious
though it might be. He'd have plenty to fall back on,
never doubt it. Plenty in reserve, that would be
Blackie. Some of these young hooligans might fancy
their chances, might go as far as tackling Webber. If
they did, the place to stand was well back from the
touchline, well clear of the field of play. Blackie would
sort out the visiting team, and there were no referees in
this business.

Still, if there was an opportunity of stirring up
dissension among the criminal ranks, Chief Inspector
Sam Hudson was not the man to miss the opportunity.
Hudson did not take a narrow or parochial view of his
allotted task in life. Some policemen think only in
terms of convictions and jail sentences, and these are
all very well in their way. Hudson's line was much
broader. The end product was what he aimed for, and
he was clear in his mind about what that end product
was. Quite simply, it was to contain, and reduce if
possible, the volume of crime in Inner London. That
was the target, and the routes to it could often be other

than the better established procedures, such as courts of law. Criminals could be persuaded to betray each other, for instance. Again, rivalry could be encouraged, to the extent where only violence could result. So many of the criminal element were vain and stupid people, that they could often be diverted into non-profitable activity within their own community, to the ultimate benefit of the community at large.

Sam Hudson never forgot a certain day when he was new to the job. He'd been on duty in an unmarked general call-car. He sat in the back, the front seats being occupied by a seasoned inspector and a sergeant ten years his senior. Radio warning of an impending affray had been received, and the car was parked in a side-street fifty yards from the entrance to a public house which was a favourite haunt of criminals. The information proved to be reliable, and soon two car-loads of men from south of the river had pulled up outside, the men disgorging themselves, and entering the place in a solid block. Soon, there were screams and shouts. Several women and one or two men came running out in panic, and disappeared.

"Sounds a bit lively," grunted the inspector. "Let's go and have a little gander."

The three officers, all in plain clothes, left the car and walked to the scene. Hudson's blood was racing, and he hoped he would acquit himself favourably in the next few minutes. The inspector walked to a window and stared in, the sergeant beside him. Hudson stood behind them, peering between their heads. The inside of the place was like a battlefield. Men were fighting, kicking, lashing out with anything that came to hand. As they watched, one man picked up a chair and smashed it over another man's head. The chair did not splinter, the way they do on the movies. The man's head split open instead, and he went down in a crumpled heap.

"Was that Dobber Martin went down?" queried the inspector in a conversational tone.

"Couldn't see properly, sir," returned the sergeant.

"Looked like old Dobber. Can always tell his ears. He won't be troubling us for a while. Oh, there's Lovely Ellis, look. Over by the bar, with a knife. Just look at his face. Wicked sod, that one. Oh, dear me, that must have hurt a bit."

Hudson was puzzled by the delay.

"Aren't we going in, sir?"

The inspector turned in astonishment, which was evidently shared by the sergeant.

"Going in, son? What for? We could get ourselves involved in something nasty in there." Then seeing the consternation on the young officer's face, he relented. "Look at it like this, son. Those men in there are all hard cases. This is not a bunch of hooligans setting about a shopkeeper. Hard nuts, every one of 'em, both sides. And what are they doing? They're carving one another up, that's what they're doing. Who loses by that, eh? They do. And who gains, do you suppose? Why, we do. The public does. Because what those bastards are doing to one another in there they could just as easily be doing to us, or some poor old lady, which is worse. We'll give 'em another five minutes, then we'll go and sort out what's left. Got the idea, constable? Right. Now then, who's going to give me a fag?"

Sam Hudson had been perturbed at first by that particularly dramatic lesson. It was brutal, unfeeling. They could have at least prevented further bloodshed, instead of leaving people to carry on. That was before he read the file on some of the people involved. After that he could readily grasp the basic reality of the inspector's decision. Since those days he had taken the lesson further, refined it, improved it. On the day of the Georgie Parks murder there were few policemen in London who could match him in the field of internecine trouble-making.

The Naughty Primrose Club had two separate existences. From noon until seven it acted as a private drinking-club, with the mixed clientèle to be found in fifty places like it. Then after one hour's break it became a strip-club, with vastly inflated prices for the mug-trade, and a few vastly inflated girls to keep them spending.

The private drinking-session was in full swing when Sam Hudson entered and made his way up to the bar.

"Afternoon, Nick. Got any Scotch from Scotland, today?"

"Ha ha, Mr Hudson," intoned Nick, unsmiling. "You know my stock is open for inspection any time you like. That'll be one pound."

"A quid? For a single Scotch?" echoed Hudson.

"The price-list is displayed for the convenience of members as required by law," he was advised. "And speaking of the law, Mr Hudson, I didn't notice you sign the book."

Hudson grinned, reaching for his pen. Cheeky sod, was Nick. Useful though, when it suited him.

"How about Georgie Parks, then?" he queried.

Nick shrugged. He'd wondered what brought the good-looking copper in.

"What about him?"

"You heard he'd been knocked off, I suppose?"

Nick took a cloth and wiped unnecessarily at the gleaming counter.

"Somebody did mention it, yeah."

I'll bet they did, thought Hudson.

"Be some fun now, I reckon," he suggested.

The proprietor narrowed his eyes.

"Fun? I don't get it."

Hudson looked confidential as he lowered his voice.

"Little bird told me Mad Maxie and that Dalston lot were thinking of taking a ride over here. Sort of retirement party for Blackie Webber. So I heard."

The trouble with Actor Hudson was, you never knew whether to believe him or not, thought Nick.

"I just run my club, and mind my own business. It's healthier."

"Don't blame you," agreed the chief inspector. "Keep out of it. That's the best way."

Ten minutes later he was inspecting a not very clean

glass in the Toadstool Club.

"Have to polish these up a bit before the party," he observed mysteriously.

"Oh?" The bartender looked interested. "Is there going to be one?"

Hudson looked wise.

"Now come on, Eddie. Everybody else in London seems to know. Posh Peter is giving it. You don't have to be cagey with me."

"Posh Peter from Hackney?"

"Only one I know," confirmed his informant. "Popping over to say goodbye to Blackie Webber. Should be quite a noisy do."

He spent the next hour roaming about the drinking-club circuit, dropping conflicting rumours about every major tearaway he could think of. The rumours all had a common basis. Someone was out to get Blackie Webber. Then, armed with a pocketful of change, he settled himself into a public telephone box, and made some anonymous calls. This time he reversed the rumours, making Blackie Webber the aggressor.

Speaking to Tony the Waiter, leader of the powerful Camberwell faction, he adopted a gruff tone.

"Tony, this is a friend. Never mind who, just marking your card. I just heard Blackie Webber saying that if you stick your face in over this side of the water you'll wind up as a motorway foundation. Know what I mean? Well, I've had enough of Blackie. Never done me no favours. Thought somebody ought to tip you off, like. Well, suit yourself."

Hudson replaced the telephone and opened his black notebook again, whistling happily.

He was having a whale of a time.

*   *   *

Blackie Webber was relieved to find a parking space outside the quiet block of flats in Kensington. Perhaps some of the people around there did go to work then, though you'd never know it from the number of drawn curtains.

They had one of these speaking arrangements where you have to tell people who you are before they release the front-door lock. Posh. This place must have been setting Georgie back a few bob. Wonder what'll the bird do now? Being a straight girl she wouldn't be earning no money worth talking about. Wonder if she'll try to touch me for a pound or two, he mused, waiting for the door to open.

Inside he went to the twin lifts, and stepped inside the right-hand one. The flat was on the third floor.

Upstairs, Olivia Marshall was trying to repair her tear-streaked face in readiness for her visitor. She'd only met the man once, and hadn't liked him at all, but that was irrelevant now. With George gone, this Webber was the only person in the whole world who could help her with certain things. Odd, the way things turn out.

Olivia was twenty-six years old, and the long awaited offspring of parents no longer young. Her

mother had been almost forty when she was born, and
her father six years older. Married for sixteen years,
they had all but despaired of ever producing a family,
and their delight at her arrival far exceeded even that
of the usual happy parents. Having been childless for
so long, they had each continued with their respective
careers, with the result that Olivia was born into near-
luxury, a state of affairs she accepted as the norm from
that time forward. She was a bright girl, although no
academic genius, and always achieved good results in
everything she set out to do. Not especially attractive
in the more conventional sense of the term, she none
the less radiated good health and intelligence,
combined with a mischievous sense of humour and an
excellent dress-flair. As a result she was always much
in demand on the social scene. It was a mark of her
popularity that other girls liked her, without too much
of the bitchiness usually inherent in such
circumstances.

"I don't know how old Liv does it," would complain
someone. "I mean, without being too much of a cow,
she's not exactly Elizabeth Taylor, is she? And yet she
draws the men like flies. Whatever it is she's got I wish
they could put it in jars. I'd buy a whole case of it."

"After me in the queue, dear. I was here first."

Attracted to the law, she could never bring herself to
doing even the minimal amount of work to pass the
necessary examinations. But she was an asset to the
small group of solicitors in Lincoln's Inn, where she
acted as the cheerful dogsbody and carved herself a

niche which suited both herself and the people she worked for.

A natural and obvious target, she led a busy social life and had her share of emotional entanglements, one or two quite serious. But she always shied away from anything permanent, preferring life to continue as it was. Until she met George Parks. George – she wished people wouldn't refer to him as Georgie – was totally unsuitable for her in every way she could think of. He was ten years older, uncouth in many ways, without any respectable employment, and on top of that he was married. True enough, he didn't live with his wife, but he was technically married, for all that. George was a villain, and she was under no illusions about that. Olivia had been to close to the seamy underside of city life to entertain any romantic notions of the Dick Turpin variety. She knew these people for what they were, and George had been a prime example. But there was such strength radiating from the man, so much animal force, that she had felt that odd weakness behind her knees which was a sure sign that something would happen between them.

Within a week they were lovers, and soon after that she gave in her notice at the little two-roomed flat where she'd been so comfortable. George had found this Kensington place, and had insisted that money was no object. Well, he'd proved that to be true, though she steadfastly closed her mind as to the source of his income. They were an ill-assorted pair, and onlookers were unanimous that they couldn't last.

Well, it had been over a year now, and they were quite content with each other. They still led their separate lives, away from the flat, but that did nothing to damage their solidarity.

And now he was gone. A great lump rose in her throat as the realisation hit her yet again. No more would he come in, whistling, tossing a fresh lobster on the silk-covered bed to annoy her. Never again would he – no. This must stop, my girl. And right now. This was not the time for maudlin recollections. There was the rest of time for all that self-indulgence.

The immediate task was to make oneself presentable for this man, Webber, and to get the mind clear as to what was to be done. Because, and she was under no illusion about this, there were certain things to be taken care of, certain arrangements to be made, and only Webber could help her. She had put herself on the wrong side of the law, and although she had never personally done anything illegal, that would not avail her if the police should come asking questions. Never mind her respectable background, parents, school, Lincoln's Inn. That would all count for nothing in the end. Indeed, it could go against her, ironically enough. How many times had she sat in court, listening to a judge saying, "These things might be understandable, even excusable, in a person of limited intelligence and from a deprived environment. But for someone like yourself, born with every advantage, well-educated, highly intelligent, there can be no room for doubt that you acted in the full awareness of what you were doing. I have no alternative but to sentence you to the maximum penalty, and you will therefore – " How many times? And what had been her own reaction on those occasions? She had agreed with the judge every time. There was no excuse for such people. People like herself. The coming conversation with Mr Webber would be one of the most important she was ever likely to have.

The doorbell chimed.

Olivia gave her hair a final pat, and went to admit her visitor.

Blackie stood there, his face composed carefully in a mask of sympathetic concern. To tell the truth, he felt a bit out of place. This girl was none of your hard-faced dollies, not the sort you usually found around and about. She was more the sort you saw out shopping in those dormitory suburbs. Still, be that as it may, she was Georgie's girl for all that.

"Good of you to come, Mr Webber. Please come in."

"Thank you. Sorry to be coming here under these circumstances."

He was rather pleased with that line, and had been rehearsing it in the lift coming up.

On Olivia, it had the hoped-for effect. He'd said that rather well, and with none of the maudlin sentiment these people usually expressed. Perhaps he wasn't going to be too bad to deal with.

She led him inside and offered a drink. But Blackie

had been rehearsing that one too.

"Not in the afternoon, thanks very much. If you happen to be making a cup of tea – "

Tea. That was a good idea. It would give her something to do, and let her get used to his presence in the place.

"It won't take a minute," she assured him.

When she went into the small kitchen, Blackie stared around. Nice. Very nice. Bit quiet for his taste. He preferred to see some bright colours about. Still, it was nice. Tidy, too. With some people, no matter how much rent they paid, they always managed to live like pigs. Nobody could say that here.

Olivia was as good as her word, and soon they were sitting opposite each other, busy with their tea cups, and wondering where to start.

"I suppose – " she began tentatively – "I suppose there isn't any more news? I mean, I've been listening to every news broadcast, but there doesn't seem to be much beyond the bare facts."

Blackie shook his head.

"Nothing that I know of. Course, I'm not exactly what you might call in the confidence of the police, you understand. They've been to see me, of course."

"Oh? What about?"

The visitor put his cup and saucer carefully down on a coffee-table.

"Well, you see, they had to be sure it wasn't me, didn't they?"

"Oh my God. I hadn't thought – "

She hadn't, either. It had never entered her head that George's own friend could have been responsible for his death. And here she was, drinking tea with the man.

Blackie smiled dismissively.

"Don't look so worried, Olivia. The first I'd heard about poor old Georgie was when that copper told me. Can't imagine who it was, and that's the truth. I suppose you haven't got any ideas, have you? I mean, I've not seen much of him these past few weeks. Has he been talking much about anybody in particular?"

Olivia shook her head, thinking.

"I suppose the police will be coming here soon."

"Not right away, I shouldn't think. But they'll find you in the end. I mean, Georgie was very cagey about where he lived. Outside of me, I can't think of anybody else who would know. But they'll know about you, the police will. You work down the Law Courts, don't you? Somewhere around there?"

"Lincoln's Inn, actually."

"Yeah, well, same thing. It won't take Bill all that long to find you, once he starts sniffing around. The firm's got your address, I suppose?"

"Oh yes. Naturally."

"Well, there you are then. What are you going to do? Move out, or what?"

He wasn't one to beat about the bush evidently. Well, if she wanted his help she was going to have to talk to him.

"I don't know," she began. "Well, yes I do know, of

course. I'll have to move eventually. When the quarter's rent is up, that is. There's no way I could afford to stay here by myself."

"No, I s'pose not. Look, Olivia, I want to help you if I can. There's two reasons, before you ask. One is for old Georgie's sake. We was good mates a long time, and I feel sort of responsible."

He looked across at her to see how she received this. Olivia nodded.

"Thank you. You said two reasons?"

"Yes. The other one is purely selfish, I'm afraid. I don't know how much Georgie talked to you about me—"

"Not at all," she interjected swiftly.

"Ah well. We had a lot to do with each other over the years. Things the police would love to know about. If Georgie's left anything lying about here which might get me into trouble, I'd like to get it sorted out before the law get their hands on it. Address books, little diaries, anything like that."

Olivia was relieved. For this man to want to help her, simply because it was in his own best interests, made the kind of sense she could understand.

"I can't think of anything," she replied. Then, seeing the disappointment on his face, she added quickly, 'But you are very welcome to look, I wouldn't want you to go away worrying. So far as I'm concerned, you can search all through George's belongings."

She wasn't going to call him Georgie, not even for Webber.

He looked pleased.

"Well, that's very handsome of you. You won't lose by it, stand on me."

Smart girl, this one. Be worth a few bob to be certain Georgie hadn't left any little molehills for nosey coppers to make mountains out of. Blackie would have liked to get up and get on with it, there and then, but the girl seemed to want to say something else. He waited.

"Mr Webber, I told you George left nothing here, and so far as you're involved, I'm sure that's true. But there is – other stuff."

Other stuff? What was she on about. What other stuff?

Olivia bit her lip before coming out with the story. Either this man was going to help, or he wasn't. She had nothing to lose. Except her freedom, she added in her bitter thoughts.

"You mentioned just now that you hadn't seen George lately. Well, the fact is, he was finding the inactivity a bit irksome. He likes to be up and doing things, as I'm sure you know as well as I do."

That was true enough. Georgie never was the boy for a lot of sitting around.

"Yeah, I know. What's he been up to, then?"

By way of reply, she stood up, saying, "I won't keep you a moment."

Olivia went into the bedroom and opened an old hat-box which she used for collecting odds and ends. She lifted out a wad of ten-pound notes and placed it on the bed. There were over six hundred pounds there, that she knew of. Then she drew out a cigarette packet, which rattled. What about the money? If Webber knew it was there he might just take it away from her. There would be nothing she could do to stop him. No, she decided. Cash was cash, and she might be very glad of it. Lifting the top mattress, she slipped the money underneath it, then went back into the living-room, and placed the cigarette packet in front of Webber.

"I don't know where those came from, or anything about them. All I know is George came home with them the night before last. He said he'd be getting rid of them today."

Blackie stared at the table, without moving. There was something valuable in that cardboard box, and he wasn't sure he wanted to know anything about it.

"Aren't you going to open it?"

Reluctantly he leaned forward and picked up the packet. Then he peered inside, said, "Oh my Gawd," and tipped the contents onto the table. Light flashed from a hundred facets as five gem-encrusted rings tumbled on top of each other.

"Been doing his Christmas shopping, has he? Little knick-knacks for the tree?" He looked at Olivia without warmth. "Where did this little lot come from, then?"

She heaved her shoulders expressively.

"I don't know. I really don't know. That's not the only lot I've seen. There have been two others that I know of in the past month. George usually kept them not more than two days, then he'd come back with a lot of money. He never told me any of the details, and I certainly didn't ask him."

More's the pity, reflected Blackie, staring at the jewellery. He was no expert and he'd be the first to admit it, but he knew the difference between real and glass. These rings were worth a thousand a time, at least, and maybe much more. But, without a fence, without a waiting market, they might just as well be glass. Worse, they were a positive danger. Nobody was going to tell Blackie that Georgie Parks had suddenly blossomed out as a Hatton Garden marketeer. This stuff had been nicked. There could be no doubt.

"Well," he said heavily. "What are you showing all this to me for, darling?"

Olivia noted the change in his tone, and flushed. He was speaking to her now in the way he probably spoke to the night-girls.

"I was hoping you'd take it away," she said stiffly, holding back the tears.

He looked doubtful. This was no time for him to get caught with stolen gear. Still, it was nice stuff, no getting away from it. Then he made up his mind.

"How much do you want for it?"

The girl looked startled.

"Want for it? I'm not asking you to buy it, Mr Webber. I'm asking you to get it off the premises."

Well, well. Still, he shouldn't be surprised really. Straight girl like this, she wouldn't know no better.

"Tell you what, let me give you a ton."

"I beg your pardon?"

What did he mean, a ton of what?

Seeing her mystification, Blackie grinned.

"A ton, dear. Hundred nicker. Pounds, that is. Don't let's call it a sale. Say you're taking the money while you're finding your feet. Getting sorted out, like."

On the point of refusing, Olivia hesitated. A refusal could suggest that she had money, and that might prompt him to search for it.

"Well," she muttered finally, "things will be a bit difficult for a few days."

"Say no more, darling."

Blackie took from his pocket a thick roll of bills, held together by elastic bands. She couldn't begin to estimate how much there was.

"There we are look, one hundred. Tell you what, old time's sake, here's an extra fifty. You was good to Georgie, I know that."

He scooped up the rings and stuffed them into a pocket.

"Now then, if I could have that little look round we was talking about?"

Olivia waved a hand.

"Please help yourself."

She hoped he wouldn't look under the mattress.

Six

Superintendent Brent was engrossed in the coroner's notes when Detective Sergeant Grayson reported back with the missing match cover.

"There we are, sir."

He placed it in front of his chief, who stared up at him.

"And how did you leave our Mr Spence? In need of a change of underwear, was he?"

Grayson smiled.

"I don't think he'll be taking anybody's matchboxes for a while."

"Did you think I was too hard on him?"

The younger man shook his head.

"No, sir, I didn't. It was a stupid thing for Albert to do, and he deserved to have his ass kicked. Think this thing might be any use, sir?"

"Hard to tell. Doesn't look like much, does it?"

The superintendent picked up the slim folder, looking at the black and gold motif on the front.

"When Sherlock Holmes or one of that lot finds a

box of matches, it's got the name of some gambling-club on the outside, and some vital witness's telephone number scribbled in it. What do I get? Black and gold squares, and half a dozen paper matches. Ever tried lighting a pipe with paper matches, Curly?"

"Can't say I have, sir."

"Well, don't bother. It can't be done."

Grayson hesitated for a moment, then spoke up.

"I did happen to notice some printing, sir, under the striking strip."

"Oh?" Brent inserted his thumbnail and bent the brown strip forwards. "So there is. Blimey, they couldn't have got it much smaller, could they?" He turned it towards the window to get maximum light. "Courtesy Gift-Matches, Mitcham, Surrey. Wonder what sort of records they keep."

Grayson had a favourite aunt who lived in Tooting. That was very close to Mitcham, and he hadn't seen her for some months.

"They usually keep sample layouts, sir, people like that. Show a prospective customer the kind of thing they've done in the past. That cover is rather distinctive. I should think they could identify it without much trouble."

"Certainly worth looking into," mused his superior.

"I could take a run over there if you like, sir. Be there and back in half a day. If you can spare me, that is."

Brent eyed him suspiciously. He could usually tell when he was being conned. They'd all like a nice drive in a comfortable car, with half a day off from the patch thrown in.

"Very good of you to offer, sergeant," he said cryptically. "Exactly what Doctor Watson would have done. But we've come up to date a bit since then. You said yourself this design is quite distinctive. Go and see if you can describe it adequately over the telephone."

The crestfallen Grayson picked up the matches and went out. Brent resumed his reading of what the coroner had found. Cause of death was confirmed as three gunshot wounds, fired at close range – tell me something I didn't know – and death had occurred within two hours prior to the discovery of the body. The blood was interesting. There was a high alcohol content, and traces of one of the soporific drugs. Still that didn't have to mean anything. Georgie Parks was no drug man, whatever else he might have been. Probably had a nasty headache and took a couple of tablets for it.

There was quite a lot to read, and Brent stuck it out faithfully, but at the end he emerged with only two usable facts. Parks had been drinking, and he'd been shot to death. Great.

No.

He pushed the papers to one side. This was not a forensic type of investigation. There were no traces under Parks' fingernails of a rare chemical supplied only to two scientists in the whole of Europe. That sort of brilliant coincidence did not come the way of working coppers. This was a straightforward shoot-em-dead killing, and he was quite certain the key lay in finding out what Parks was doing in Cookstone. It was as he'd said to Sergeant Grayson that morning, and nothing had happened since to change his mind. And how long did it take these people outside to complete a few simple enquiries?

Striding to the door, he flung it open and bellowed, "Sergeant."

The desk sergeant came in on the run.

"Sir?"

"Haven't you uniformed people come up with the answers on those public transport enquiries yet?"

"Oh yes, sir. Being typed now."

"Typed?" he echoed, his voice rising. "Typed? What do you think this is, sergeant, a parking offence? This is a murder investigation, sergeant. Murder. A man is dead. I'm supposed to be finding out who did it. I'm sitting here, starved for information, while your constables are practising their typing. Time is of the essence, man. I must know everything the very second the information becomes available, not when your people feel like it. Now then, get those investigating officers in here. P.F.Q."

"Sir."

The sergeant hurried out. Brent groaned in frustration. It wasn't going to get him anywhere shouting at the local talent. They weren't geared up for this kind of stuff. The last gunshot incident on the patch had been three years earlier, when a poacher tripped and shot himself in the leg.

There was nervous tapping at the door, and two red-faced constables came smartly in, standing to attention. Brent glowered at them.

"Sorry to interrupt your typing," he growled, "but I'm a very inquisitive sort of man. I don't want to wait until your book is published, got me? I want to hear it now. Who covered the railway station?"

The taller of the two inclined his head.

"That was me, sir." He pulled a notebook from his pocket and opened it. "At eleven-fifty a.m. I interviewed the superintendent – "

His voice died away as Brent shook an impatient hand.

"Just tell me, son. This isn't the bleeding magistrate's court."

"Very well, sir, the answer is no. There were no unidentified strangers who arrived at the station yesterday."

"You're quite sure about that?" demanded Brent.

"Not much room for doubt, sir," was the confident reply. "The people on duty yesterday were all local men. They know their regulars, and were able to give good descriptions of the few people they didn't know. It didn't take long to establish that no one like the deceased passed through the railway station."

"Then why does it take so long for me to be told?"

"I pursued my enquiries outside the station, sir. There were two men among the strangers, and I wanted to know more about them, just to be on the safe side."

"And – "

"Turned out to be visiting estate agents," the constable advised. "But it took a while to track down the taxi-driver who carried them."

"And that's all. You're quite satisfied?"

"Yes, sir."

"All right." Brent turned to the other officer. "Now then, what about the buses?"

"There are only four a day, sir. Two morning, two afternoon. It's a pay-as-you-enter service, so the drivers get a good look at every passenger."

"Did you talk to all the drivers? You've been quick if you have."

The constable grinned.

"Only one man involved, sir. That's his day's shift, doing the round trip four times. He's very knowledgeable about all his passengers, treats them all like members of the family. There is no way he could forget picking up someone like the dead man."

"H'm. So the answer is a lemon. All right, don't let me keep you from your typewriters."

They went out, passing Sergeant Grayson, who was waiting for his chief to be free. Brent made a face as he entered.

"How do you stand it down here, Curly?"

Grayson was almost caught off guard, but remembered in time that this was not a casual remark in the canteen. This was the fearsome Superintendent Brent, from County, and anything he said in an unguarded moment might be retained for the future.

"It's not bad, sir, once you pick up the pace."

"Slow down to it, you mean," grumbled the super. "You got something for me?"

"Yes, sir, the matches. Been talking to the senior sales manager of the company. Struck me as a very knowledgeable, go-ahead sort of man. Good witness, if you know what I mean."

"Thank Gawd there's someone we can rely on. Does he remember our little exhibit?"

Grayson nodded.

"Almost certainly. Of course, he wouldn't take an oath on the strength of a telephone call. He'd want to see the cover itself – "

Still looking for a ride out, decided Brent, interrupting. "Then you can send him a coloured photograph. Carry on."

Well, no harm in trying, thought the sergeant, undismayed.

"Shall be done, sir. Mr Drysdale thinks it's a special order, one thousand covers at a time, twice a year, for a place called the Arena. It's in Denman Street, in the smoke."

Brent nodded with satisfaction.

"Good. We might have learned something from your Mr Drysdale. What do we know about this Arena, anything? Something to do with boxing, is it?"

The young sergeant shook his head.

"No, sir. It's to do with gladiators. The Arena is a gay bar, sir. Gladiators are very big with that – er –

fraternity.''

Grayson's tone was expressionless, but his face spoke volumes. The superintendent chuckled.

"You ought to see your face. Well, what do we make of all this then, eh? Has Georgie Parks gone queer on us, do you suppose?"

He was really talking to himself, and not expecting an answer. Grayson stared at the photograph on the wall of the victorious police team in the local six-a-side football competition.

"Never heard anyone suggest that Parks was fruity, but then, we'd have to ask the London crowd about that. I'd always heard he was quite a ladies' man. In fact, it's got him in trouble in the past. Not with us, but with the ladies. Can't imagine him being bent."

Grayson cleared his throat.

"Still, you do hear of these cases, sir. Man who's been married for years, family and everything, suddenly decides he's been doing it all wrong. Case only last week in the *People*. Not all that uncommon."

Brent listened without enthusiasm. He didn't like the idea of Parks being one of that lot. Still, there could be profit in this, so far as the case was concerned.

"I don't want to start any hares running around this nick," he said pointedly, "but obviously we can't ignore this development. Let's take a new look at things. Suppose Georgie Parks was a homosexual and nobody knew anything about it. Just suppose. Now, is it possible that he came down here to visit somebody with the same inclinations? I would say the answer to that is yes. Yes, it is possible. So, where does that get us? What's the score around here in that direction?"

"Not very much, sir," was the reply. "But I'd like a few minutes to look into that angle before I commit myself."

"Fair enough. But – " and Brent pointed a warning finger – "I don't want any rumours starting out there. You keep your cards in your hand, sergeant. And I mean it."

"Very good, sir."

When he'd gone Brent picked up the matchcover and looked at it with renewed interest.

Gladiators, eh?

We live and learn.

*   *   *

Keith Firbank pulled up outside his little cottage and climbed thankfully out of the car. In the past thirty-six hours he had had precisely two and a half hours' sleep. One didn't have to be a very good doctor to know that a man can't keep up that sort of pace for very long. At that moment his ambition was to have a bath, put some kind of snack meal together, and be in bed by nine o'clock.

Unlocking the front door, he stepped inside, bending to pick up the usual thick pile of mail. Most of it would be junk, he knew, the usual collection of offers, free samples and so forth, pressed on every general practitioner in a seemingly endless stream. Throwing it all on the hall table, he kicked the door shut, sighed with tired pleasure, and went to pour himself a drink. A few minutes later, glass in hand, he wandered back into the hall, staring at the pile of mail.

Better just flick through it, he decided. Could just be something important buried somewhere in there. Most of it was as he had feared, but there were three worthy of early attention. About to open the first one, he changed his mind, slipping all three into the side pocket of his jacket. There could be something in any one of them that would set him in motion, required him to do something, and that he most certainly did not want. If he wasn't careful he'd find himself rushing off somewhere, or getting busy on the telephone. There would be no bath, no early night. And why? Because he had been unable to resist the impulse to open an envelope.

Well, it wasn't going to happen tonight. Those envelopes could remain in his pocket, at least until he'd had his meal. Better still, what he would do would be to change into his pyjamas after his bath and be ready for bed as he ate. That way, no matter what the mail had to say, he would be a man actually on his way to bed when he read it.

Good idea.

He went off to run his bath, splashing in some of that pine stuff his mother had produced for him the previous Christmas. Normally, he was not a man given to what he called 'smellies', but this particular product did give out a most invigorating smell.

The telephone rang.

"Blast."

Turning off the taps, he went to the small living-room where the telephone rested on the drinking-cabinet.

"Hallo," he grunted.

"Dr Firbank?"

A woman's voice, clear and pleasant.

"This is Nancy Parfitt."

"Oh, hello, Mrs Parfitt. What can I do for you?"

It was the blonde from the surgery, the one with the long legs. Keith Firbank had a good idea of what he could do for her, and if she kept this up she was going to get it. Country practice or no.

"It's the Wilson boy, young Joseph. You have his file with you, and his mother made an appointment for tomorrow morning while you were out on your rounds. Dr Patterson is seeing him first thing, and there are some papers to be added to the file. I wondered if it would be an awful nuisance if I popped in to collect the file?"

"Please don't put yourself to that trouble, Mrs Parfitt – " he began.

"Oh, it's no trouble. I have to come out your way in any case."

He thought quickly.

"I see. Well, it's very good of you. I'm just on my way out, actually. What I'll do, I'll leave the file on the front porch. It'll be quite safe there, and I'm sure no no one will pinch it."

There was a pause. Nancy Parfitt had evidently not forseen this possibility.

"Are you going out at this instant?" she queried. "I can be there in five minutes."

But he had adjusted his thinking now.

"Already started the car," he assured her cheerfully. "I only came back in because the telephone was ringing."

The disappointment in her voice came over clearly.

"Do you know, I've got a better idea," she exclaimed. "I've only just remembered that I have to come into the village early tomorrow morning anyway. Over half an hour before surgery. So what I'll do, I'll just knock on your door as I go by then, if that's all right?"

Firbank gave up. She had been too quick for him with this second salvo. Anyway, not much could come of it at eight o'clock in the morning, surely.

"What a concientious girl you are," he said pleasantly. "Very well, then. I'll see you in the morning. Good night, Mrs – "

"Please call me Nancy," she begged. "Everyone else does."

"Well of course. Good night, Nancy."

"Good night, Keith."

He grinned to himself as he put down the telephone. Perhaps he ought to find out a bit more about this girl.

Well, that would have to wait till morning. The thing now was to get back to the evening programme. Where were we?

Bath. First job.

It was not until an hour and a half later that he sat down in his most comfortable chair, bathed, fed, and at peace with the world. The unopened letters were waiting on the coffee-table, and now at last he was ready to deal with them. He could identify two of them from the envelopes, but the third was a mystery. It was obviously personal, the name and address having been scrawled by hand, but the writing was unfamiliar. There was a London West End postmark. Abandoning further speculation, he ripped open the envelope and lifted out some pink sheets which smelled faintly of perfume. Mystified, he unfolded the pages, and suddenly felt quite sick. Sick and outraged.

What right had these bloody people got to invade his privacy? How the devil had they found out his address, anyway? He wanted nothing more to do with them, and here they were, actually sending things to his private house.

Thoroughly disturbed, he threw the papers on the floor, got to his feet, and went to pour himself a large drink. Downing it savagely, he glared at the offending papers. They had scattered untidily, but the top fold of one had fallen in such a way that the elaborate gothic script of the printed heading was clear to see. It consisted of one word.

Arena.

Seven

Blackie Webber drove expertly through the busy London traffic. He was uneasy, as any man has a right to be when he has a pocketful of valuable jewellery which doesn't belong to him. Blackie had no ambition to find himself charged with being 'caught in possession'. Besides, it might not end there, for all he knew. Some poor sod of a jeweller, or householder, might have got his head blown off in the process of being separated from his property. That would make the charge much worse than possession. No, any way you.looked at it, the sooner he got these baubles off his person the better it would be all round.

There was a multi-storey, twenty-four-hour car-park in a side-street at the back of Woolfridges. Blackie locked up the car, and walked down to a rear entrance of the famous department store. Advancing briskly through the leather goods displays, he headed for the basement steps, looking quickly around before descending. At the foot of the steps, he identified himself to the uniformed attendant, who released the electronic catch on the steel-grilled door leading to the strong-boxes.

"Please go through, Mr Cutler."

'Mr Cutler' smiled pleasantly as he passed into the privacy of the locked vault. Here were the rows of steel units, identical in outward appearance but each with a lock unique to itself. Blackie had held a key for years, but quite often made no use of it for long periods. It was handy if a man found himself suddenly with a lot of cash he couldn't explain. You couldn't exactly put it in your Post Office savings book, or any other channels open to people who acquired their money by honest means. No, the strong-box was the only place for money, and it would do very well for this stuff he was carrying at the moment.

There was no one else in the vault at the time, and he was unobserved as he took one last quick look at the flashing gems before depositing them in the anonymous darkness of the box.

Emerging from the store, he had a new spring to his step, as befitted a man who had recently had a load lifted from his mind. What he needed now was a place where he could have a quiet think. Somewhere with a telephone. He didn't particularly want to do any driving at that moment. The car was quite safe where it was. Leave it, he decided.

"Taxi."

Ten minutes later he was sitting at a quiet table in a small but exclusive drinking-club behind Piccadilly Circus. This was not one of your tourist traps, and

indeed was not even available to the public at large.
When the place said 'Members Only', it meant it, and
the only path to membership was by carefully vetted
application. You had to be somebody, reflected
Blackie with smug satisfaction, before they even let
you in the place. This was the place where big money
was taken for granted. You needed more than that to
acquire the coveted member's pass. You needed
status, and that was something he had striven for.
What was it the Yanks called it? Respect, that was it.
You had to have respect.

He signalled to Louie, the red-jacketed barman,
who came over to his table.

"Yes, Mr Webber."

Mr Webber, you see. None of your Blackies which
might get you in some places. Class, that was.

"Got a visitor coming in, Louie. A Mr Stone. Tell
the people at reception I'm expecting him, would
you?"

"Certainly, Mr Webber."

Louie melted away. Blackie lit a fresh cigarette and
settled quietly back to wait. Ollie Stone was the
manager of one of the betting-shops and he would be
busy putting the required information together at that
moment. Busy wondering, too. He didn't normally see
Blackie more than a few times a year. Georgie Parks
was the minder, when it came to the shops. But old
Georgie wouldn't be available any more, and that left
Blackie with two choices. He could either put his
managers on their own responsibility, or he could do

Parks' job himself while he was looking for a successor.

Well, it was early days to be making up his mind about that. This coming chat with Stone was for quite different reasons. Not that Stone needed to know that. Blackie had learned from experience that it was a great mistake to let too many people have too much information. All they needed was exactly enough to enable them to carry out that part of the job, whatever it was. Outside of that, tell 'em nothing. What they didn't know wouldn't hurt them. Or you, for that matter.

The man he was expecting arrived suddenly in the doorway, staring around the place. Blackie half-raised an arm from the table, and the visitor's face broke into a wide smile as he advanced over the carpet. When it came to running the shop, balancing the books, sorting out unruly customers, all that side of life, Ollie Stone was a whizzbang. You could not wish for a better man. He would look after your interests, your money and your staff. Where Ollie fell down, was in the department of looking after himself. A short, square man, running to premature baldness, Ollie was one of those people who simply could not present himself. You could put him in a brand-new shirt and tie, suit just back from the cleaners, and brush him from head to foot. Get him immaculate. Within five minutes he would contrive to look as though he had narrowly escaped death in a train wreck, or some natural disaster. The suit would be rumpled, the shirt awry, and the shoes would defy description. In any discussion on the world's ten worst-dressed men, the only points at issue would be the identity of the remaining nine. The number one spot was reserved exclusively, and indisputably, for one Oliver Stone of London.

He stood now at Blackie's table, beaming and sweating profusely. His shirt collar was undone, the tie a ragged string. Extra portions of him were available for inspection where his shirt buttons had come undone just below his spreading stomach. Blackie sighed inwardly, hoping that Louie would not find him guilty by association.

"Sit down, Ollie, sit down. You want a drink?"

Ollie shook his head, pulling out a chair.

"You know me, Blackie. One glass of sherry every Christmas."

That was another great point in his favour. You didn't have to worry about him spending all his time in the pubs. Ollie took a handful of peanuts. Some went on the table, some on the floor. The rest went into his mouth, and he scrunched noisily away.

The new arrival waited patiently. He wasn't going to open the bidding. It was Blackie's call, and Blackie was the boss. You could easily say the wrong thing, and find yourself in trouble.

"We'll have to keep our voices down," began Blackie softly. "Wouldn't do to disturb people, am I right?"

"Right," agreed Ollie vigorously. Then seeing the

pained expression on Webber's face, he repeated the word much more quietly. "Right."

"You've heard this terrible news about Georgie, I suppose?"

Ollie swallowed quickly. He'd heard all right. He'd also wondered whether he might know anyone who'd been present when Parks had left this world.

"Terrible," he agreed, nodding.

"There'll have to be one or two changes, Ollie. I'm thinking that over at the moment. Don't want to rush things."

"I see, I see."

Ollie didn't see at all, but it wouldn't do to say so. His wasn't the only betting-shop in the Webber chain, and he was wondering why he had been sent for. Where were the other managers? Or perhaps Blackie would be seeing each in turn. He would have been even more intrigued if he'd known the real reason he was being quizzed. Ollie's betting-shop was just around the corner from Hatton Garden, the centre of London's precious stone trade.

"You got that list I asked for?"

In their telephone conversation earlier Blackie had required him to prepare a list of all those punters who were in debt to the shop for amounts of one hundred pounds or more. As a request, it wasn't so unusual. It was just the timing that was odd, what with Georgie being bumped off and everything.

Ollie handed over his piece of paper. There were only eight names listed.

"This is all?"

"That's all, Blackie. I don't like people to go on too long. They can get forgetful, like, know what I mean? So I gee 'em up every couple of days."

Blackie nodded, studying the names. There was one down there he'd been hoping to see, but it would never do for Ollie Stone to know what he was up to.

"I'll just hang onto this for now."

He slipped the paper into his pocket. Ollie nodded unhappily.

"They're all right, you know, Blackie. I mean they'll pay all right. I know all them people."

Blackie knew what was going through his mind. He had visions of Blackie, or some of his strong-arm squad, leaning on the customers. That would be bad for business. Bad for Ollie's image, too. Either he was in charge of the shop or he wasn't. Seeing the anxiety on his manager's face, Blackie was tempted to put his mind at rest. Then, just as quickly, he changed his mind. Let him sweat. It would do him good, do 'em all good, to be reminded that they weren't in charge of anything at all, unless he said so.

"That's all I want for now," he said brusquely. "Ta-ta, Ollie."

"I've brought the takings balance with me – " began Ollie, but he was waved down.

"I'm too busy now. Perhaps tomorrow. See you."

The dismissal was final. Ollie got up from his chair, managed to grab it before it hit the floor, and made his untidy way out. Blackie watched him go, then took out the paper again, smoothing it with his banana-clump hands.

There was the name of the man he wanted to see.·

'Tommy Halt. One hundred Forty Pounds' read the entry.

He hadn't seen little Tommy for ages.

"Louie," he called, "is the telephone room vacant?"

*　　*　　*

When Chief Inspector Sam Hudson was out on what the department called general surveillance, and what he called roaming about, he made a practice of telephoning the office at hourly intervals. On this occasion he was at Green Park when he made contact with the duty officer.

"Hudson," he reported. "Anything for me?"

"Yes, sir. There was a telephone request for you to call a Superintendent Brent as soon as convenient. I've got the number here, shall I read it out?"

Hudson thought for a moment, then decided against it. Whatever Brent wanted, it might involve reference to files, or to other people.

"No," he replied. "I'll come into the office and do it. Tell my people I'm on my way, will you? Give 'em a chance to sort out anything I should be looking at."

"Will do, sir."

One of the difficulties Hudson found with his present job was keeping up to date with the paper-work. This was always a major item in every policeman's life, but with Serious Crime he'd found it worse than most. When the squad had first been formed, a few years ago, it had taken people a while to settle into the new system. So far as all existing police authorities were concerned, crime was crime, and they were equipped to deal with it. They had no intention of handing over their jurisdiction to some gang of know-alls in London. Fortunately for the future of the new team, the Home Secretary of the day had taken a personal interest, and a few heads had been banged. The original trickle of paper had become a steady stream. Then, when the value of the Serious Crime squad became self-evident, both with its success rate and its diplomatic dealings with regional forces, the stream became a flood. The members of the squad sometimes felt that any felony greater than a bicycle theft was marked automatically 'copy to S.C.S.'.

There was always an enormous pile of reports to be sifted through, and that day was no exception, as Hudson found when he saw his desk.

"Great sky above," he muttered, staring at the in-tray. "Do you know what I'm going to do one day, sergeant?"

The waiting sergeant grinned expectantly. You could always rely on a gag from Sam Hudson.

"What's that, sir?"

"I'm going to set up a little team of my own. Analyse all this stuff, as to origin and referring officer. Especially the referring officer. I'll take a small bet with you some of these blokes are major shareholders in a paper-mill.''

"Could be, sir. Er, there's three there that I think you ought to know about. I've marked them all and put them on top of the pile.''

"Right, thanks. Just leave me to get on with it, eh?''

He settled down behind the desk and picked up the first case. It was interesting enough, a large country house robbery in the Lake District, but he couldn't get involved, somehow. His mind kept going to the Parks murder and all the mayhem he hoped would result from it. One thing was certain. If there wasn't some trouble among the tearaways in the next few days, nobody could say he hadn't tried.

Pulling the telephone towards him he picked up the receiver.

"I'm supposed to return a call to a Superintendent Brent,'' he announced. "Have you got the number there? Right. Get it for me, please.''

The telephone rang almost at once. That was quick.

"Hudson,'' he said, without preamble. "Is that Superintendent Brent? What? Oh, I see. You're calling me. Wait a minute, let's start again. I've just put a call in, and I thought this was it. I am Chief Inspector Hudson. Who are you, please?''

"My name is Walters, Insurance Protection Partners Limited.''

"What can I do for you, Mr Walters?''

The caller hesitated.

"It's rather a delicate matter, actually. Not something I'm anxious to discuss over the telephone.

I'd rather come into your office and talk to you, if that's convenient."

Hudson looked at his watch.

"Or I could call in and see you," he suggested, "if you're not too far away."

"Well, no. If you don't mind, I would prefer to come there. We like to keep a very low profile, you see. It wouldn't do for people to see a senior police officer coming in. People think we're rather harmless, as things stand, and that's an image we'd like to preserve. I'm not far away, actually, only Victoria Street. Would it be possible to see you now? Or, say, give me ten minutes to get there?"

Hudson made a face, but could not quickly think up a refusal.

"Yes, all right. This won't take too long, will it?"

"Five minutes should do it."

"All right, Mr Walters. See you soon."

He put down the phone. Insurance Protection Partners Ltd? Could mean a lot. Could mean nothing at all. Well, he'd soon know.

When the telephone rang again, he took no chances.

"Chief Inspector Hudson," he said formally.

"Sam? This if Mike Brent."

So it was going to be Sam and Mike after all. Mr Brent must be after something.

"Found something on George's person that might help a bit. Matchcover. Comes from a place called the Arena. It's in Denman Street, they tell me. Do you know it?"

Hudson was mildly affronted. Of course he knew it. It was his job to know the West End. Never been inside, though.

"I know where it is," he replied cagily.

"One of my chaps thinks it's gay. Is that right?"

"Yes. Very well-behaved place, though. Expensive, I believe."

There was a moment's silence while Brent digested this new information.

"Really? Well, I stand to be corrected, and of course I haven't got your first-hand knowledge of the West End lot, but it strikes me as a bit odd that our deceased friend should be going there."

Hudson nodded, as though the caller could see him.

"I agree. Nothing wrong with Georgie. If anything, he's been a bit too normal with the birds, our Georgie. Matter of fact, I've got a man out at this very moment tracking down some woman he's living with. He hasn't been near his home in weeks. You're quite right, Mike, this is a rum 'un. Can't fit Georgie into that bent picture at all. Would it help if I had a stroll around there, sort of size the place up?"

At the other end Brent smiled with satisfaction. It was the offer he'd been hoping for.

"If you can fit it in," he said guardedly. "You know how it is. If I come up making a big official request out of it, and all that, it makes a big hoo-ha out of what might be nothing at all. It's different for you. You're on your own patch. People expect you to turn up all over the place."

All of which was perfectly true, mused Hudson.

"Tell you what, then. I'll have a little butcher's later on tonight. Nothing fancy, just a little sniff. If it looks like getting naughty, I'll leave it out, and come back to you. That suit you?"

It was a fair, not to say generous, offer. Brent was delighted.

"I shall owe you one. You'll let me know how it goes, then?"

"Tonight, if it looks promising. Otherwise, first thing tomorrow."

Hudson was thoughtful after their conversation. Pushing a buzzer he summoned the young sergeant back to the office.

"Any of our drop-outs around, sergeant?"

The sergeant looked at his watch.

"Mostly out on the street by now, sir. I might be able to catch a straggler."

"Off you go, then. Tell him I won't keep him a minute."

The sergeant was back almost at once.

"Just caught the last one at the back door. He's waiting outside."

"Wheel him in, then."

Leaning back against the open door, the sergeant nodded his head.

"Chief Inspector will see you now. This is Sergeant Green, sir."

"All right. Shut the door, please."

Hudson stared at the apparition before him. Greasy

long hair dangling in front of an unshaven face. Beads draped from the neck over a filthy buckskin jacket. The inevitable stained jeans, and a pair of once-nice loafers. You could stand this object in the dock in any magistrates' court in the country and make up your own charge. The verdict would be 'guilty' before the clerk had finished reading it out.

"Lucky we caught you, sergeant," began Hudson. "Need a bit of information."

"Will it take long, sir? I'm due on a bust in an hour's time."

"Two minutes. Denman Street, the Arena. Some sort of gay centre. Do you know much about it?"

Green shook his head.

"Not much to know, sir. Place has only been operating a few weeks. But it's strictly high-class, very respectable, if you can use that term. You won't find any runaway boys chained to the beds, or any of that rubbish. None of your public lavatory squad. These are all professional people, and some of the boys on the squad have been a bit surprised at some of the faces they've seen going in there."

"I see." Hudson had expected something of the kind, but unfortunately it did not match with what he wanted to hear. "Can you join at the door, do you know? You know what I mean, temporary member five quid, that sort of thing?"

The beads rattled as Green shook his head.

"No way. They have to know you before they'll even let you complete an application form. Then it's a week or more before you get through the door. The place has been set up to protect the members, and they're not going to run any risks."

Hudson tried to keep his tone matter-of-fact as he asked his next question.

"Have we got our own member, sergeant?"

The face opposite drained of all expression.

"I'm afraid I have no knowledge of the assignments of other officers, sir. Any requests for such information would have to be channelled through – "

He stopped as the chief inspector chuckled, and waved him down.

"All right, you scruffy layabout, I know the rules. Anyway, you've been very helpful. Thanks very much, and good luck with your bust."

"You're welcome, sir."

Segeant Green disappeared away to his half-lit world, but Hudson was not to be free of visitors.

"There's a Mr Walters outside, sir. Says he made an appointment."

Walters? Oh yes, the insurance bloke.

"So he did, sergeant. Show him in, please."

Eight

The sudden pinging sound made Olivia Marshall jump. Really, she was going to have to get a grip on her nerves. Still, what was she to expect after what had happened to poor George? Large tears welled unheeded, and rolled down her face. Pull yourself together, girl, she snapped.

Crossing to the voice-panel she called out, "Who is it?"

"Miss Olivia Marshall?"

A man's voice, that was not familiar.

"What do you want?"

"I'm a police officer, madam. I'd like a few words with you, please."

They had been quicker than she expected. It was inevitable they would find her eventually, of course. Still, thank heaven that jewellery was gone. She had that man Webber to thank for that, regardless of his motives. But she didn't want to see the police. Not now, not today. In two or three days' time, perhaps. Give her a chance to pull herself together.

"Are you alone?" she demanded.

"Why yes, madam," was the puzzled reply.

"Have you got a warrant?"

"No, madam. This is a purely routine enquiry – " began the voice.

"Then you're not coming up here," she stated definitely. "Anyone can say he's a policeman. You go away, and come back with another officer, and then I'll let you in. Not before."

There was a sigh of exasperation from the small square grill.

"Very well, madam. If you insist. I shall be back within the hour."

The firmness in his tone was unmistakable. He was certainly the genuine article, and he would equally certainly be back.

The interruption had been the very thing Olivia needed. Since Blackie Webber had left she had been in a state of indecision. Unable to accept the reality, the very finality of George's death. Unable to visualise her new position as a woman suddenly without any roots at all. There'd been little enough with George, but somehow that hadn't seemed to matter so long as he was there. Without him, there was nothing, and for the first time in her life she felt afraid. It was all most uncharacteristic of her. She was always the one who knew what to do, and did it. This vapid creature, wandering about the flat, feeling George's suits, sitting on the bed and staring, this wasn't Olivia at all. She was the girl who accomplished things, the one who showed the way.

The unseen policeman had restored the balance.

She was now a threatened person. Not in the physical sense, although she couldn't be one hundred per cent certain about even that, she admitted. Someone had murdered poor George after all, and that seemed senseless enough, so there could be no absolute guarantee of her safety. But it was not in that area she felt threatened. She had seen what the media made of women like herself, women who had widespread notoriety thrust on them overnight. It didn't take a great deal of imagination to visualise the headlines that would be splashed in certain newspapers about her. George Parks had been a criminal, and she had no illusions about that, even though he kept her unaware of the details. But the newspapers would not be satisfied with such a milk-and-water description. Anyone could be called a criminal, and the public were not interested in small fry. George Parks, by the very fact of his murder, would become automatically a famous London gangster, a kingpin of crime, and all the rest of the cheap-jack soubriquets. And Olivia knew perfectly well what that would make her. Not that she gave a damn, personally. But others would suffer. Her parents in particular. Her married brother and his family. Even the law firm where she worked would not be immune.

Well, before they could photograph her, misquote her, and follow her around, they would have to locate her. At present, that wasn't too difficult, as the police had already demonstrated. She might go free for the rest of the day, but the Press would inevitably trace her by the morning. The policeman had said he would be back within the hour, but she would be wrong to accept the upper limit. She was probably safe for about twenty minutes, and no more.

Crossing to the white leather telephone, she pressed at the buttons. The receiver at the other end was lifted on the second ring, and a man's voice said, "Hallo."

"Is that you, Ronald?"

"Olivia?" He sounded excited. "What the devil's going on, Liv? We had a police officer asking after you today. Wanted your home address. Had to give it to him, of course. Tried to phone you, but nobody answered."

That was true. The telephone had rung a number of times while she'd been giving vent to her grief.

"Sorry about that," she apologised. "I must have slipped out to the shops."

"What's it all about, anyway? Have you had a car accident or something?"

The poor man was clearly quite concerned about her, as they all would be. It was that kind of firm. It was fortunate that Olivia had already decided on her story, because the temptation to tell Ronald the whole truth was great

"No, it's rather more serious than that, I'm afraid," she said in a trembling voice. "The fact is, I've had a bereavement. Someone very close to me. I'm afraid I shall have to be away for a few days."

"Oh dear." He was all warm sympathy. "Is there

anything we can do to help?''

"I'm afraid not, but thank you for offering. I'm sorry the notice is so short, but my desk isn't too much of a mess. That Chalmers file, by the way, is with old Mr Ellis. You'll be needing that. Otherwise, things aren't too bad. I don't know quite when I'll be back.''

"Now don't start worrying. Take all the time you need. Oh, and your salary will continue, naturally. Where would you like it sent? Have you an address?''

"No,'' she replied quickly. "No, nothing is fixed yet. If you'd be kind enough to transfer it into my bank I'll be grateful.''

And that, after some mutual expression of goodwill, took care of the office problem. Earlier that day, with no clear idea of why she was doing it, Olivia had begun several times to pack her bags. Each time, the hopelessness of everything, the bleakness of the future, would overcome her, and she would abandon the job. Now, with her new-found resolve, it was a different matter. She had never been a dithery packer, taking hours to decide whether a certain pair of shoes would be required, and her long-standing efficiency stood her in good stead now.

Ten minutes later, the two dark green leather suitcases stood waiting. Thank heaven she didn't have to worry about cash. What with the six hundred George had left behind and the money his friend Webber had pressed on her, she had seven hundred and fifty pounds in ready cash, apart from the few pounds already in her purse.

Walking to the bed, she lifted the mattress and took out the roll of bills. At that moment she tried to visualise Blackie Webber's face if he were able to see what she was doing. They'd joked about it, while he was searching for anything George had left which might incriminate him.

"Well, that's about it, Olivia, and thanks for the co-operation. Nothing here that will harm either of us. I haven't looked under the bed, of course."

It was a joke, and she found herself able to smile.

"There's nothing under there except George's burglary kit. A black mask, a jemmy, and a sack with 'Swag' stencilled on it."

"Oh yes, like in the kid's comics," he chuckled. "Yes, I used to read all them. Well, thanks again, darling. I'll be on my way."

Tucking the cash deep inside her largest handbag, she took one last look around. No, there was nothing else she needed. It was even possible she'd never come back here again. Strange, what a transient business real life was. People were supposed to plan any major change for months ahead. Yet here she was, all packed and ready to leave at a few hours' notice.

Now, then. Taxi.

With her hand on the telephone, she drew it back. No, that wouldn't do. They probably kept records, and the driver would be able to remember where he took her. She'd walk as far as the High Street and hail a cab that was cruising. That way she'd be just another face.

Letting herself out of the flat, she closed the door, then, in a moment of quick sadness, rested her face against the panels. The moment passed, and she picked up her bags and walked to the lift.

Five minutes after she left a police car pulled up at the front of the building, and two officers climbed out.

"Perhaps the lady will be a bit more obliging this time," snapped one.

At that precise moment a taxi-driver swung into the kerb where a well-dressed young woman stood with two pieces of luggage. Very nice-looking fare, this.

"Where to, miss?"

"Heathrow, please. Terminal One."

* * *

Nancy Parfitt looked at her watch for the hundredth time, and drummed her fingers impatiently on the table-top. She'd known this golf-club do of Derek's would be a wash-out. They always were. The pattern was so well-established as to be insulting. A lot of formality, and clustering round the women for the early part of the evening. Then, as though by a signal, the men would suddenly be missing, clustered in noisy groups in the various private bars, and extremely resistant to being pulled away.

"Shan't be more than a few minutes."

"Order yourself another drink, old girl."

"Be as quick as I can. Can't be rude to the chaps."

How many times had she heard it all? Even the

dialogue was the same. Some of the women didn't seem to mind much. They had become inured to it over the years, and far too many used their own brand of compensation by applying themselves to their glasses.

Well, she wasn't going to go down that road. Three gins was her limit. After that, she felt either ill or fell asleep, and that was no way to spend an evening. Even such an evening as this.

A roar of laughter from the bar brought her head round. That must have been a particularly coarse one, judging by Derek's expression of bloated joy. What a pig the man was, once he got in with his cronies. He even looked like a pig, with his clean-scrubbed, fattish face.

"Derek's enjoying himself," her neighbour commented. "Looks as though he is, anyway."

Nancy smiled tightly.

"Do you know what he looks like, to me? An advertisement for somebody's pork sausages. Just like a well-nurtured pig. Don't you agree?"

Far from feeling discomfited, the other woman took the question very seriously, sipping at her seventh large brandy while she pondered.

"Bit of a farmyard all round, isn't it, dear? We know what we are, don't we? I mean, we're just the prize cows, herded off into a corner out of harm's way. Now let's think. If Derek's a pig, what'll we do with my Hubert?" Turning her head towards the laughing men, she nodded wisely. "I've been thinking of a bull but that would be flattering him a bit, don't you think? He's too scrawny altogether. Scraggy, perhaps. How about a rooster? Do you think Hubert looks like a rooster?''

"A bit."

Nancy was afraid she might start to scream if something didn't happen. It was only nine o'clock, and she knew there was no prospect of getting Derek away before eleven at the earliest. It was all so bloody unfair. There were plenty of men who would be very pleased at the prospect of having her all to themselves for an evening. They'd think of better things to do than sit around in this god-forsaken club. And as for leaving her to fend for herself – well.

Take Doctor Firbank, for one. She hadn't missed that quick electricity between them earlier in the day. True enough, he'd switched it off almost immediately, but that didn't conceal its existence. Nancy was nobody's fool, and she could understand his point of view completely. Handsome young doctor, unattached, and so forth. They had to be more careful than most people. Just the same, there was no denying the interest she'd aroused in him, nor its nature. And in that nasty old white coat they all had to wear, as well. If she could manage to stir him up while she was wrapped up in that boring old garment, what would happen if he got a look at her now? She'd worn her provocative blue tonight deliberately as a male-catcher. It had been her hope that the men might pay her a little too much attention, and Derek would be compelled to stay close to her, to protect his own interests. At first, it had seemed her plan might work, but once the wine started to flow she was no competition for the bar-tender. What a waste it all was.

Well. It didn't have to be, she decided.

Nancy stood up.

"I'm going home," she announced. "Bit of a headache, and I've got a long day tomorrow."

One or two of the other women put on sympathetic faces. She gathered up the little silver purse and went across to tell Derek she was leaving. He made one or two noises of protest. They were small noises, and it was quickly agreed that she would take the car. One of his dear, close friends would give him a lift home later.

"Make it as late as you like," she told him, "but please don't wake me up."

Several pairs of male eyes watched the tall lithe figure as she walked gracefully away. More than one of Derek's dear, close friends wondered if he ought not to offer to see her home, see if there was anything he could do. But at the vital moment someone shouted, "Whose turn is it?" and attention was brought back to the proper business of the meeting, the momentary distraction forgotten.

The brandy-drinking woman leaned confidentially towards her neighbour.

"Jonathon now, more of an elderly sheep, wouldn't you say?"

"What?"

* * *

Doctor Keith Firbank was about ready for bed. It was warm and comfortable in the small living-room, and he had drunk two large nightcaps. After a very long stretch without sleep, and following on top of a bath and a meal, these had induced a most pleasant langour all through him. There'd be no disturbance tonight, either from the police roster or the emergency service at the practice. He was definitely off duty, and that meant the greater part of ten hours' sleep. What a lovely prospect.

Stretched out in his most comfortable chair, he congratulated himself on his forward planning. Here he was, all ready to go to bed, and all ready for bed as well, having changed earlier into his pyjamas and dressing-gown. That was foresight, he thought smugly. Some people had to leave the warmth downstairs, then go up to a cold bedroom and take all their clothes off. Next thing they knew, the cold had woken them up, and they'd lost the advantage of getting sleepy in the first place. That was some people. Other people, people with a certain flair in these matters, foresaw all these problems and dealt with them in advance. This second category of people, the ones one could categorise as night thinkers – yes, that was rather good, that, the night thinkers – they didn't fall into that trap. Not they.

There was only one problem these more enlightened

people had to grapple with. Did they, or did they not, want just one more spot of the good stuff before retiring? Let's see, he'd had two already. Well, no, hang on a minute. That didn't mean two measures, the way it would in a public house. This wasn't a pub, and there was no measure here, other than his eye. No, in pub terms he'd probably had four at least, perhaps more. Well now, Dr Firbank, that being the case, do you really feel that another is justified?

On the other hand, there would be no harm done. There was no motor car to be driven, no patients to be considered. This was entirely a matter between himself and his conscience.

What was that noise? Sounded like someone at the front door, but it couldn't be. Not at twenty-past nine at night. Must be the wind. Let's see, if one of my visual measures equals say two and a half pub measures, that would mean –

There it was again. Dammit, it was the door.

Reluctantly he pushed against the arms of the chair and stood upright, gathering the dressing-gown round him and yanking hard at the belt. Who the devil could be dragging a man out of his chair in the middle of the night? Well, not quite the middle of the night. Still, late enough. Late enough.

Sliding back the flimsy chain, he opened the front door.

Good Lord. It was the blonde, the one with the legs. Mrs – er. Call me Nancy. That was it.

"Oh, hello, Nancy. You'll have to excuse my

appearance. I was just – er – "

Nancy Parfitt smiled, and her eyes glittered. So you were, my boy. Well, my little intrusion need not delay that in any way. Exactly what I had in mind, in fact.

"Sorry to disturb you, Keith," she said, and waited.

Firbank stood there grinning fatuously, and feeling rather foolish. Then he realised he was keeping the girl on the step. There was practically no top on that dress. She'd freeze to death.

"Oh, sorry. Do come in. Will this take very long? It's that file you want, isn't it? It's around here somewhere."

She stepped inside, brushing lightly against him, and letting the expensive perfume do its work. Then she stood in the tiny hallway, like a small girl awaiting instructions.

Keith closed the door, his mind working nineteen to the dozen by this time. Nancy Parfitt – that was it, Parfitt – hadn't come knocking on his door for the simple purpose of collecting a file. Not dressed like that, she hadn't. The point was, what was he going to do about it? It would be a comparatively simple matter had circumstances been otherwise. Enjoyable, too, he had no doubt about that. But he was still a village G.P. and she was still –

"What a dear little place. I've often wondered what it was like inside. All women wonder how bachelors live, you know."

"Yes, I expect they do. Look, Nancy, won't you come into the living-room? It's not very tidy, I'm afraid. Could I get you a drink? I was just thinking of having one."

She didn't want a drink, having reached her limit earlier at the golf club.

"That would be lovely," she assured him. "But only to my own prescription. Doctors aren't the only ones who have prescriptions, you know. Mine is, one part gin, nine parts tonic water, and I warn you, I can taste it at once if the measures are not exact."

He grinned. This one was evidently no boozer, anyway.

"Well, come through."

She followed him into the living-room, wondering what he meant by untidy. Some men lived in the most appalling fashion, as she recalled from her student days. But not this one. The room was really very presentable. It lacked the final touches, which were the sure signs of a woman's hand, but that was to be expected. On the whole, he was quite a tidy man.

He came up to her, holding a tumbler.

"Nine to one," he bowed gravely.

"Cheers."

She sipped at it, and he was as good as his word. There was almost no trace of gin. Even so, it was not her intention to finish it. She had no intention of making herself ill, which was one possible outcome if she did. As to falling asleep, well, if she were to do that, it would be for very different reasons.

The hand with which he had extended the glass was still hanging uselessly in the air between them. She looked up into his eyes, and her own were soft with unspoken permissions. All he had to do was to take half a pace forward.

He swallowed nervously.

"The file." He almost shouted the word. "I've a fancy it might be with some of my other papers upstairs. I won't be a minute."

Nancy was mildly disappointed by this setback, but she'd seen the look on his face, felt the hunger coming from him. This wasn't the surgery, and there was no desk between them now. When he came back to her she would be in control of the situation, not him. And everything was going to be exactly as she had envisaged.

She wondered, with delicious anticipation, what he would be like. Not clumsy, anyway. She'd watched those deft hands at work selecting instruments. Rough? He might be. You could never be sure with these self-contained ones. Well, dear, no point in too much speculation. We'll know in about – what, ten minutes? No, give him fifteen. That would give her plenty of time to be home in bed before Derek got back.

Hallo, what was that on the floor, behind the chair. He'd dropped some papers there and would probably spend hours looking for them. Setting down her glass, she leaned over the back of the chair and picked up the pink sheets.

Arena. What did that mean, she wondered. Sounded like some kind of sports centre. Idly she

began to scan the typed pages, not making much sense
of it, at first.

Oh, my God.

It couldn't be true. Not Keith Firbank. He couldn't
possibly be one of these – these people. Then why did
they write to him? Of course, it explained him.
Explained why he'd never raised one word of gossip in
the village. Plenty of those cows had tried for him, she
knew that. This must be the answer. He wasn't a man
at all. Not a real man. He was –

Oh, God.

Nancy felt suddenly sick. She might as well have
drunk his bloody gin, after all.

Away. She must get away. There was no way she
could spend another second in this house, alone with
this – this pervert. Who could tell what he might do to
her.

Turning, she ran from the room, out into the hall,
and flung open the front door.

Firbank had just begun to descend the stairs,
clutching the file which was their joint flimsy excuse
for her presence. He heard the front door open.

"Nancy?"

He ran down the rest of the way, only to hear a car
door slam and the sudden roar of the engine.

"Nancy?"

He dashed out of the door onto the pathway, waving
the file. Nancy Parfitt hadn't even bothered to switch
on the car lights.

Scratching his head in bewilderment, Keith went

slowly back inside, closing the door. What had got into the girl? He'd already decided, while he was upstairs, that he would be a fool to turn down the opportunity she was presenting to him on a plate. He had even worked out one or two little manoeuvres in his head. They may not work, and she might even turn him down at the last minute. It was a game some of them played. But under no circumstances would she run away. What on earth could have upset her like that? Because obviously something had. Unless she'd suddenly developed cold feet. That was a possibility.

Still worrying at it, he walked back into the living-room. The first thing he saw was the crumpled pink paper, lying now in the middle of the floor.

That was it.

She'd found that stuff from those Arena people, and read it.

He sat down heavily, all sleep gone.

What a mess.

Nine

That same evening, Blackie Webber sat in the baroque splendour of an East End saloon bar. He'd summoned up three of his team for the outing. It would not be sensible for an obviously affluent man like himself to be parading the streets alone, and apart from that, it would be lowering to his status not to have a few of his lackeys on view. You never knew who might be watching, and there was always some young lunatic with ambitions about moving up in the world.

There were only two of his men with him at that moment, the third having been dispatched on an errand.

"Eddie's back," grunted the man next to him.

Eddie approached the table.

"Little Tommy's on his way in," he reported in a low voice.

"Good, good." Blackie looked at his companions. "Tell you what, why don't you all go and have a nice game of bar billiards? Won't keep you long."

They melted away, and Blackie sat smoking and

watching the door. Within minutes it was pushed open and a short, ferrety-faced man scuttled in, looking anxiously around. He saw Webber at once, blanched, swallowed mightily, and went up to the waiting man, smiling and bobbing.

"Evening, Blackie. You was looking for me?"

This could only mean trouble, very bad trouble, and Tommy Holt was beside himself with fear. He couldn't understand it either. Mr Stone said he was all right for the money for a day or two. Still, what Mr Stone said didn't matter now. This was the big cheese himself, Blackie Webber, no less. It was what he said that mattered.

Blackie nodded, without warmth.

"Sit down, Tommy. You're quite a stranger. How're things at the Garden these days? Busy, are they?"

Tommy sat down on the edge of a chair, ready to bolt if necessary.

"Always plenty going on. You know how it is."

"Yes, yes. I was having a chat with Ollie Stone only today, matter of fact. You seem to be getting behind with the rent, Tommy."

He lowered his eyelids, inspecting the hapless little man through the slits that remained.

"I talked to Mr Stone about it," squeaked Tommy. "He said he wouldn't worry for a day or two."

"Did he? Did he really?"

Leaning forward, Blackie crushed his cigarette slowly and deliberately in the ash-tray. Tommy Holt's eyes popped as he watched the careful destruction.

"How would you feel about doing something for me? I'd appreciate that, and it might go a long way towards helping you out."

Holt heard the question with amazement. What could he possibly do for the great Blackie Webber?

"Anything," he replied hurriedly. "Only too pleased to help you if I can. You don't need to ask."

"That's what I thought you'd say," Blackie nodded. "After all, what are friends for if we can't help each other in times of trouble, eh?"

Holt didn't care for the word 'trouble' but he agreed vigorously.

"Pull your chair in a bit. This is confidential."

When Blackie spoke again, his voice was little more than a whisper.

"You know me, Tommy, every man to his trade, I always say. I never poke my nose into the glass business, and the glass boys don't meddle with me. Everybody's happy, am I right?"

Tommy did not care for the sound of that at all. The precious stone people were specialists, and just as deadly in their own field as Blackie Webber was in his. He didn't want to get mixed up in any wars. At the moment he didn't have to commit himself, merely agree.

"Quite right, Blackie."

"Yes. Well now, I know you keep your eyes and ears open, Tommy. Don't get me wrong, I know you never repeat nothing. That's why people trust you. But you do hear things. I mean, if something was going on you'd know all about it, eh?"

"I might. Depends on what it is."

"Realise that, realise that. Now, we all know there's a lot of shenanigans goes on with the jewels. Stuff coming in without bothering the Customs, stuff going missing, the police don't get told. A lot of work done over the phone, on street corners and that. I mean, that's all common knowledge, that is."

Was it? Tommy Holt wasn't aware that it was all that common.

"Lot of people know about that," he agreed.

"Course they do. But they don't know the details, they don't know the names. Don't know how much money changes hands, all that sort of thing. Only the real Garden people know all that, eh, Tommy?"

Tommy Holt was becoming more unhappy by the minute. He was on a loser here, and no mistake. If Blackie Webber was going to start muscling in on the Garden crowd there'd be bloody murder. And if they suspected that little Tommy Holt had opened any doors he'd be the first one on the slab.

"They don't tell me much," he protested miserably. "I'm sort of an errand boy, really. They're what you might call secretive. Know what I mean? If I ever do hear things I'm supposed to keep stumm."

"Course you are," endorsed Blackie heartily. "Course you are, son, and I respect you for it. Nobody respects a talkative man. But just supposing, only supposing mind, a man heard of a few stones that had got lost somehow. What would you advise a man like that to do? Who would he talk to?"

Tommy chewed at his thumbnail. So that was it. Blackie had collared some stolen gear. Not like him at all.

"It would depend," he muttered. "Quality of the gear, that's one thing. How much of it, that's important. An then, of course, no disrespect, you understand, it would depend how naughty it was."

Blackie knew what he meant by the last part. That was the same with any stuff that went missing. If there'd been any nasties while it was still going missing, like somebody getting accidentally killed with a shotgun, then the price went down, Right down. The trouble was, he didn't know the history of the stuff locked up in his strong-box. Not that he was disposed to confide any further in Tommy Holt. He was an eyes and ears man, and that's all he was.

"I'd have to discuss that with a number one man," he said crisply. "No need for you to bother your head with a lot of details. Just tell me who I talk to."

Tommy hesitated.

"Could be awkward," he hedged. "They're very sensitive, you know."

"Sensitive? We're all sensitive. Take me, now. I'd be proper upset if those hooligans on the bar billiards was to take you outside and smash all your fingers. Really put me off me supper, that would. Anyway, there's no need for all that. You're going to tell me, aren't you, son?"

"What about the money? The hundred and forty quid?" blurted Holt.

"One good turn deserves another, I always say. You put me on to the right party, and if it all goes well we'll tear up your little piece of paper, eh? Can't say fairer than that. Mind you," and Blackie pointed a stern finger, "it'll have to go well, son."

"All right. It's a swap."

Blackie beamed magnanimously.

"That's a good, lad. Now then, this name."

* * *

"Mr Walters, sir."

Chief Inspector Sam Hudson rose from his desk to greet the visitor.

Mr Walters was above medium height, and thin as a rail, in an immaculate dark blue suit. They shook hands, and Hudson waved him to a chair.

"Very good of you to see me at such short notice," began Walters.

"You picked a time when I happened to be in the office," replied the policeman. "Most of the time I'm out there."

He gestured towards the window. Walters inclined his head.

"Then I must capitalise on my good fortune and not waste your time. Have you heard of the firm, chief inspector?"

Hudson looked at the card which had been laid before him.

"Insurance Protection Partners Ltd., " he read out. "Well, I may have. To be frank, there are so many names that sound alike in the insurance world – "

He left the sentence unfinished.

"Just so. Well, we are a little unusual. We don't, for a start, sell any insurance. Ah, I see I have your attention."

"You have indeed," confirmed Hudson. "What do you do, then?"

"We provide a service to most of the leading houses. It's really very simple. We collect information, on claims over a certain sum, and collate it. Let me go a little further and confine myself to the loss areas. By that I mean burglary, theft, fire and so forth. There are people in the world, it pains me to repeat, who are constantly on the look-out for means of defrauding the insurers."

Hudson grinned.

"I had noticed that," he confirmed.

"I'm sure you have. So have all the companies. In fact, they employ people, specialists, who are really quite excellent at detecting – um – irregularities. It occurred to my chief a few years ago that much of this expertise was lost to the world at large by being confined within one company. It seemed to him that if the more unusual cases were given wider publicity, then other firms would be able to gain. No individual firm could be expected to set up a special department merely to keep the competitors informed, but they might be prepared to pass the details on to one common source. This source could then look after the interests of all concerned, and at comparatively little cost to each individual organisation.''

Sam Hudson was beginning to wonder where all this was leading.

"Sounds reasonable,'' he said cheerfully. "And hence your company.''

"Quite so.'' Mr Walters was coming to the point now. "Recently, there have been four cases of burglary, involving the removal of quantities of jewellery, cash and other disposables. They each seemed to be quite routine, so long as one may use such an expression in relation to criminal activities. They took place within a thirty-mile radius of London, and three different police forces were involved. That's the first point. The second point is, that three different insurance companies carried the cover. Thus, with different policemen carrying out the investigation, and different companies receiving the claims – ''

"There would be no one to point out similarities,'' Hudson chimed in. "If it were not for the existence of a certain organisation where all the information was put together. Yes, I see. You are beginning to interest me considerably, Mr Walters.''

Walters beamed with satisfaction.

"That was my hope, I must confess. You see, our cross-referencing systems for purposes of data comparison have produced some interesting – or perhaps I should say worrying – facts.'' He held up his hand and proceeded to use his fingers to number off. "One. In each case there was minimum damage to the property. The thief or thieves knew exactly where the valuables were kept. Two, in the first three cases the valuables were returned within a matter of days, in consideration of a percentage of the insured value – "

"Let me interrupt you a moment," Hudson chipped in. "You're telling me that the thieves claimed the reward money themselves? I don't think I like this, Mr Walters."

Walters looked pained.

"That is not the precise position," he contradicted stiffly. "What happens is that some entirely blameless third party is given knowledge of the whereabouts of the stolen property, on condition that he negotiates the payment of the specified amount. It is a system deprecated by everyone, chief inspector, but consider the point of view of the covering company. They have a straight choice. Either they must pay, say, ten thousand pounds to the insured person, or one thousand to some – um – other quarter. When you look at it in those terms, you see, they really have very little choice."

As a police officer, Hudson did not like it at all. But as a pragmatist, he could see the force of the argument.

"You said this only happened with three of your four cases," he queried. "What happened to the other one?"

"Yes, yes, I was just coming to that. The most recent case was only two nights ago – I have all the details here in my brief-case – and as usual a telephone approach was made concerning the return of the goods. An arrangement was come to, and the transaction should have taken place today. It did not happen, chief inspector.''

''The thieves didn't keep their end of the bargain?''

''Quite so. However, we can return to that. If I may just continue with the points of similarity. Three, in each case there was no proper security system on the premises. Just the usual trivial precautions, locked windows and so forth. Nothing that would deter an experienced man. Four,'' and now his eyes gleamed, ''Four, chief inspector, is my most salient point. All these householders had made enquiries about such systems in the past year or so. They had all received estimates of cost, and for one reason or other decided not to proceed.''

Sam Hudson wasn't certain whether he was missing the point, or whether the precise Mr Walters hadn't made it yet.

''Forgive me if I seem dense,'' he apologised, ''but all that tells me is that getting one of these alarm systems put in is an expensive business. What's the point?''

''The point, sir, is, that it was the same company who quoted in all four cases. Now, as you appreciate, in order to give such a quotation it is necessary to have access to the house, examine all windows and doors, and so forth. To obtain, in other words, a detailed knowledge of the lay-out.''

The chief inspector bowed his head.

''Yes, I know a little about it. But what you're

suggesting is that this firm is robbing its own customers. Bit far-fetched, isn't it? I mean, these people are subject to a lot of safeguards, police scrutiny and so forth."

"Heavens yes, and believe me, the company concerned is above reproach. But you are incorrect in one particular, if I may make the point. The people who have been robbed are not customers. They were, at one stage, potential customers, and that is very different. That means that all the details of the property were recorded, but no further action taken. It would be possible, would it not, for one unscrupulous individual to keep a note of such cases, wait a reasonable length of time, and then put that knowledge to criminal use?"

"H'm."

Hudson didn't like it, but he could see the sense of the argument.

"It's possible, I suppose," he agreed grudgingly. "Even so, it could be any one of a dozen people in the firm, I imagine?"

"No." Walters' denial was quite definite. "No, I think not. Because, in each case, it was the same representative who made the initial visit and inspection, and carried out the quotation."

That certainly sounded a bit more promising.

"And you think it might be worth our while to have a chat with this man?"

Walters spread his hands.

"The thought had crossed my mind."

Hudson tapped a pencil on the desk-top, thinking.

"If you'll forgive my saying so, Mr Walters, you seem to have taken your time about bringing all this to police attention. Could I ask why?"

The visitor did not appear remotely offended.

"Years ago, when I was new to the insurance world, I used to go rushing to the police with the most remarkable pieces of deduction, on the very flimsiest of evidence. A great deal of the time, I was wrong. Again, even if there was something to be said for my theories, it was quite often totally insufficient for any police activity. One learns not to rush, Mr Hudson, when one has acquired the requisite number of bruises."

Hudson grinned at the truth of that.

"But you don't think you're going to get bruised on this one?"

"No. There is one last point I wish to put to you. When the last – um – transaction took place, I had already made a preliminary deduction along these lines, and I contrived to be able to observe the principals involved. After the money had been paid over to, as I said previously, an impeccable third party, I kept watch on that office. A man arrived within minutes, and left with a small hold-all. I had a good look at this man, who was by no means an office worker, or indeed any kind of respectable person. I tried to see where he went, but, alas, I'm no private investigator."

The policeman contrived to keep impatience from his face. Bleeding amateurs. More trouble than they are worth.

"You lost him, eh? Well, it's easily done. Lost a few myself."

"I'm afraid so. Today, I was all prepared for another surveillance, but as I said earlier, the contact was not made. I think I know why." Walters opened his brief-case, and took out a folded newspaper. He spread it out in front of the watching Hudson, tapping at a photograph on the front page. "I could be mistaken, but I don't think so. This man, here, is the man who collected the proceeds from the last burglary.'

Hudson stared at the smiling face of Georgie Parks. This could explain a lot of things. If Parks had been mixed up in all these burglaries, he could have been killed for the proceeds. Or for the money he'd already collected from the earlier jobs. One Superintendent Michael Brent was going to be interested in all this. Still, he demurred, this wasn't Parks' cup of tea at all. He'd never gone in for this kind of work. And certainly Blackie Webber wouldn't. Too risky, by half. Webber wasn't the man to risk his freedom, and Parks worked for Webber. On the other hand, Webber had been very inactive lately, and it was possible that Parks had become bored, looked around for something to keep him busy. Our Georgie always liked to be up and doing. Always used to, rather.

"This kind of crime is not at all characteristic of this chap," he said slowly. "I must say, I like the way you've put your case, Mr Walters, but at the same time, if I start the ball rolling an awful lot of coppers will be doing an awful lot of work. They're not going to love me if this turns out to be a bad 'un. You're quite positive in your identification, are you? There's no room for doubt in your mind?"

He watched the other man's face closely while he waited for the important answer. There was no hesitation either of voice or expression.

"No doubt whatsoever. I saw him from half a dozen different angles. That is the man, beyond question."

Hudson clapped his hand down on the table, making everything rattle.

"All right, Mr Walters, I'm going to go for it. Lord help me if I'm wrong. Now then, let's start at the beginning. This representative from the security company. You have his name and address?"

"Oh yes. All ready for you."

Mr Walters handed over a piece of paper.

"Hugo Latimer," read Hudson. "Sounds like a writer or something. Well, the name means nothing to me. I'll have C.R.O. run it through the mill, be on the safe side. Shepherd's Bush address. Well, I'll get one of my men to check at that end, find out a bit about our customer. What's this?"

Another sheet of paper was being held out to him.

"That is a description of the property missing from the latest burglary. Might help you with identification, if it happens to turn up."

"Right, yes. Good. Well, Mr Walters, for better or for worse, I think we're going to be busy for a day or two."

Ten

Police Constable Tom Hibbert was reflecting on the unpredictability of life. His life in particular. Less than twenty-four hours previously he had been pedalling quietly around his peaceful beat, with only the occasional rabbit for company. Then, quite suddenly, he had been pitch-forked into a whirl of activity. Police activity, medical activity, and news media. And, as suddenly as the storm had erupted, it had subsided. Not merely calmed down, but gone to ground. No more was he confronted with cameras, men with notebooks, hawk-faced superiors. The tide had receded, and died away, and there was nothing now on the surface of his familiar beat to give any sign of what had happened, except for one patch of flattened grass by the side of a quiet lane.

Funny old business, he was thinking, as he pushed effortlessly at the pedals. You'd think at least he would be on the telly, one of those up-to-the-minute news programmes.

"And now we are proud to bring you the man in

person, the police officer who was first on the scene of this dastardly crime.''

That sort of thing. He would have worn his new suit; Ethel would have wanted him to do that. Of course, uniform is always dignified. They might prefer him to wear his uniform.

He didn't even hear the car, until it was on him. It came roaring round the bend much too fast, and on the wrong side of the road. The car didn't even have lights on. He noticed that almost at the moment of impact. Instinctively, he had pulled his front wheel sideways, in a hopeless attempt to avert the inevitable. Nancy Parfitt yanked on the steering-wheel in the same moment, with the result that the front bumper hit the rear of the bicycle, sending Tom Hibbert sailing into the air, to land in a heap against the hedgerow.

Nancy came running back up the road from where the car had screeched to a halt. There were tears streaming down her face as she bent over the inert figure. Oh, God, it was Mr Hibbert.

Telephone. She must get to a telephone. There was a cottage a hundred yards back. They would be able to help.

Oh, God. What a mess.

The people in the cottage were exactly the right couple for such an emergency. They sat Nancy down and made her drink some brandy. While the woman looked after her, the man got busy on the telephone All the emergency services swung into action, and less than an hour later Nancy Parfitt was sitting in the little police station, waiting for Sergeant Grayson to return from the hospital. They'd given her a cup of tea, and she sat holding it, like a dutiful child, and making no move to drink it.

The duty sergeant watched her, stone-faced. He was an old colleague of Tom Hibbert's, and he felt no warmth towards the woman who'd run him down. She'd been drinking, into the bargain, you could smell it on her. Pity those people at the cottage had given her that brandy. They only meant it for the best of course, but what they'd done in effect had been to ensure there could be no charge made of driving under the influence. Shame, that was. Still, there was a whole bookful of other charges to be thought about, and him and Curly Grayson were just the boys to dig 'em up.

He knew her sort well enough. Out boozing half the night at the golf club, then thought they could go screaming around the lanes with impunity. Well, we'd see about that, my girl.

Mind you, she did look ill. Sort of shocked, you might say. And well she bloody might. Not as shocked as poor old Tom at that minute, you may rely on that.

Joe Grayson suddenly walked in. The duty man raised his eyebrows in enquiry.

"He's all right. Luck of the devil, old Tom. Be right as rain in twenty-four hours. Nothing broken, thank the Lord." Grayson turned unfriendly eyes on the girl. "You're very lucky, Mrs Parfitt. Lucky I'm not charging you with manslaughter right this minute. Mrs Parfitt, are you listening to me?"

She gave no sign that she had heard him. She simply sat, holding the enamel mug, and staring at the whitewashed wall. Grayson looked at his colleague.

"Has she been talking to you?"

"Not a word. Been like that half an hour or more."

"H'm." Grayson scratched at his chin thoughtfully. Some kind of shock. Reaction, that's what it was. He'd seen it before. He'd better get a doctor in on this. Could be nasty, and he didn't want to face any accusations of failure to bring in medical advice. "We'd better send for a doctor."

"Not Firbank."

The woman had sprung to her feet at the word 'doctor', enamel mug splashing to the floor. The two sergeants looked at each other.

"We ought to have a doctor in, Mrs Parfitt," explained Grayson gently.

"Not Firbank," she repeated flatly. "I won't have that pervert touching me."

She was definitely going out of her mind, this one. Pervert? Dr Firbank? He might not be Grayson's favourite G.P., but there was certainly nothing abnormal about him. Wait a minute. Her car had been coming from the general direction of the Firbank house. She was all dressed up for a party, or had been before her dress got torn. Grayson was suddenly intrigued. Walking up to where she was standing, he took her quietly by the arm.

"That's an odd word to use about Doctor Firbank," he encouraged. "You did say pervert, didn't you? Why, Mrs Parfitt?"

"Ask him," she replied tonelessly. "Ask him and his dirty friends at that place."

Grayson was more interested than ever. What was this he was about to hear? Some kind of sex orgy going on in the locality? The woman had lost interest again, by the expressionless look on her face.

"What place would that be, Mrs Parfitt? The place where we find these people? Local, is it? Dr Firbank's house, perhaps?"

Nancy Parfitt shook her head suddenly, and her eyes widened. Where on earth was she, and what was going on? Oh yes, the accident. Poor Mr Hibbert, and then those nice people made her drink that brandy. She should not have let them do that. She knew perfectly well she'd already had her quota. Her feet were wet. Why was that? Looking down, she saw the brown pool all around her. Looked like tea, or something. And what was this good-looking young policeman on about? Something going on at Keith Firbank's house. He was wrong there. There might have been something, if she hadn't found out about him in time. Ugh.

"Mrs Parfitt?"

"What?"

"You made a face, and said 'Ugh'. Why? What's going on at Dr Firbank's house? And who are these friends?"

Nancy sat down very quickly. God, this was an awful mess all round. If only she could think clearly, get her thoughts in some sort of proper sequence.

"Were they still there when you left?" pressed Grayson.

"No," she blurted out. "I mean, there wasn't anybody there. Not so far as I know."

"But you were there?" he insisted.

Well, no harm in that, as it happened.

"Yes. For a moment. I had to collect a file from him."

"I see." Grayson was puzzled by the whole conversation, but highly relieved that she seemed to have snapped out of her trance, or whatever it was. "Mrs Parfitt, you just made some rather serious allegations against Dr Firbank in front of two police officers. I'm afraid I must ask you to explain. First, you referred to his friends, or more accurately his dirty friends, and then you said he was alone. Well, Mrs Parfitt?"

Nancy bit her lip, and wished she could remember exactly what she had said.

"I didn't mean they were there," she said nervously. "I read about them. In the magazine."

"The magazine?" Grayson looked across at his colleague, who hunched his shoulders.

"Yes. It was on the floor. He'd dropped it there. I picked it up, out of curiosity. It was addressed to him all right."

Grayson's face cleared slightly. So Firbank read

dirty magazines. Well, you could never tell about that. Funny, this young woman getting so upset about it.

"What was it called, this magazine?"

"Can't remember. Something Roman, I think. Stadium. Circus. One of those words."

Detective Sergeant Grayson took a deep breath. It was absolutely ridiculous, of course, and there couldn't possibly be any connection. Still, he'd have to ask.

"Arena?" he suggested casually. "Could it have been Arena?"

She nodded at once.

"Absolutely," she confirmed. "That was it. Arena."

"There's no doubt in your mind?"

"None at all. It's printed on pink paper."

Grayson whistled through his teeth.

"Better get this place tidied up, Cliff. I'm going to bring Superintendent Brent down here."

* * *

"Come in."

Sam Hudson looked up from his desk at the man who walked in.

"Ah yes. What have you got?"

"You told me to check on this Hugo Latimer, sir. He's gone, I'm afraid. Scarpered."

Hudson clucked with annoyance.

"Sod it. Any more?"

"Not as bad as it sounds, sir. He doesn't seem to

have left the country, or anything. He's been planning to move for some weeks. The locals seem to like him, and he's told them about it."

"Then where's he gone?"

His informant looked uncomfortable.

"Shan't know that until the morning, sir. The person who will have his new address is the manageress of the flats. She's spending the night with her married sister, and nobody – "

" – knows where the sister lives," finished his chief sourly. "Well, I haven't any justification for setting off the alarm bells. We'll just have to wait. A few hours won't hurt. You look as if you're going to tell me something else."

"Well, sir, I didn't like to come back with nothing at all. Thought I'd just take a little look around the flat."

"Oh, yes? Gave you a key, did they?"

The visitor grinned.

"I didn't think you'd want me to bother people, sir. It was only an ordinary mortice and tenon lock."

"Tut-tut," reproved Hudson. "That's breaking and entering, that is, my lad. What will you get up to next? And?"

"Clean as a whistle. There'll be plenty of dabs, of course, but I didn't hang around for that sort of thing. One interesting fact, though. I think our Mr Latimer might be a queer. There was a bit of mail on the mat, circulars mostly. But one of them was from a high-class gay club in Denman Street. The Arena. Thought you'd be interested."

The Arena. What a coincidence. Mike Brent had been asking about the very same place in connection with Georgie Parks' murder. And now, here it was again, cropping up in this other matter. Small world. He'd better go and have a chat with these people.

"Got a car outside?"

"Yes, sir. I didn't check it in, because I thought you might want to practise your lisp."

"You are awful, Julian. You drive."

It took fifteen minutes to reach their destination in the choked evening traffic. The car came to a halt under a sign reading 'No Parking at any time. Maximum Penalty imposed'. Hudson climbed out, looking at the sign and the two broad yellow lines which disappeared under the wheels.

"All right for us," he grunted. "What about all those other poor sods trying to find a space?"

The driver shrugged.

"They'll have to join the force then, won't they?"

The entrance to the Arena was unobtrusive, to say the least. There was a black-painted door, with an illuminated sign no bigger than a post-card. The door was evidently soundproof, and there was no handle, nothing to grasp on that impassive surface. Hudson pressed the small buzzer.

A voice spoke at once, from a small grill above the door.

"Your identity, please?"

"Police," announced Hudson cheerfully. "Open up."

The door opened immediately, and a large man regarded the visitors from thoughtful, granite eyes. The two men held out their I.D. cards.

"Ah, yes," he said, in a pleasant voice. "Please come in, gentlemen."

They followed him inside, and the door closed behind them with a hiss. He lead them to a small room marked 'Manager' and knocked.

"Two gentlemen from the police, Mr Varden."

If the police officers had been expecting any garish clothes, or any of the more outrageous outward signs, they were disappointed. Mr Varden looked like any city gent, in his sober dark blue suit and tie. He smiled at them without warmth.

"And what can I do for you this evening?" he asked in a deep, well-modulated tone.

"Want to get some information about some of your members," Hudson could see no point in beating about the bush.

Mr Varden looked pained.

"This is a private club Mr – er – "

"Hudson. And it's not Mr. It's Chief Inspector Serious Crime Squad."

"Ah. Serious Crime Squad. I doubt if we can help you very much. The people here are vetted most carefully. Their private lives are quite impeccable, I do assure you."

Sam Hudson knew something of the club's standing. It wasn't one of those places where you could start leaning on people. Not this early in the game, at least.

"Quite appreciate your position, Mr Varden," he returned smoothly. "And in fact, our little bit of business has no connection with the Arena, so far as we know. Like to keep your name out of it altogether, if we possibly can. It's hardly the club's fault what people get up to in their spare time, is it? I have no wish to make life difficult for you. All I want is information. There are two people involved."

"I see. Well, I am sure you appreciate this is a very private place, and really rather new. We haven't yet had much opportunity to know everyone. Perhaps if we tried the names?"

"Start with Hugo Latimer."

There was a small card-index sliding drawer in the desk. The manager flicked through it and found the card.

"Yes. We have Mr Latimer. What about him?"

"What sort of chap is he?"

"Let me think a moment. Latimer. Oh yes, I remember him now. Young, about thirty. Good-looking man, well-dressed. I believe he's some kind of an insurance executive. But you must know all that?"

"Don't know much about him at all," denied Hudson. "Bit of a mystery man, you see. Anything else?"

Varden hesitated before speaking again.

"That's interesting that you should call him a mystery man," he said finally. "Quite frankly he's a bit of a mystery to me, too. I can't think why he bothered to join. Mr Latimer does not seem to share our – um – special interests."

The two policemen looked at each other.

"Then how did he come to be a member? I thought you were very careful about vetting applications."

"Oh, we are, we are," assured Varden hastily. "Let me explain. We took over the lease from an earlier club, the Blue Pussycat, a very different place altogether. Since we were new, and in order to avoid any unnecessary embarrassment, we circularised all the people on their mailing-list. This served two purposes. It would save a lot of people coming here under the wrong impression, for one thing. Also, it drew attention to our existence if anyone wanted to become a member under the new arrangements. Mr Latimer was one of these, I believe." His eyes dropped again to the card. "No, no, I'm wrong there. He came along as a guest. The other man decided not to proceed with his interest, but Mr Latimer filled in the form."

Hudson's interest quickened.

"This other man, do you know who he was?"

Varden shook his head.

"Sorry, no. We have this little system of recording on the cards how someone came to us in the first place. This green tick here means that it was connected with the old mailing-list, but that doesn't tell us who he was with."

That was a disappointment, but the chief inspector showed no sign on his face.

"Have you got a card there for Parks? George Parks."

The manager's violinist fingers flicked at the cards again.

"No, sorry. I have no one by that name."

A small square photograph appeared suddenly under his nose.

"Ever seen that man before?"

Varden stared at the picture of George Parks, taken during one of his many police interrogations. It was a good likeness, and recognition was instant.

"Oh, him." He said brusquely. "Yes, he's been here once or twice. And now that I see his face, I remember that he's been signed in by Mr Latimer. Quite frankly he is not the type of person we want here. He has made rude remarks more than once. I spoke to Mr Latimer about it myself."

"Ah." Hudson leaned over the desk. "When was that, Mr Varden?"

Varden made a face.

"Oh, ten days ago. A fortnight perhaps."

"But when did you last see either of them?"

The man behind the desk considered.

"Mr Latimer was in here last night, or was it the night before? One of the two. We don't keep a check on our members' movements, you know."

"This is very important, Mr Varden. This is a private club, and I'm sure you're very careful about the licensing laws. Before your people can be served a drink they have to sign the register. Could someone check it, please?"

"Of course. As you say, we are most careful.

Meticulous, in fact. I will go and check at once."

"Thank you."

Varden proved to be taller then expected when he stood up. He went out of the room, and Hudson muttered to his companion, "What do you think of this lot?"

At the same time he was looking at Latimer's card. It bore the same address in Shepherd's Bush as the one he'd recently vacated.

"Seems like a legitimate set-up to me, sir. But why would this Latimer come here, if he's straight?"

"Don't know," grunted his chief. "Unless it was for the sole purpose of meeting our Georgie. I mean, it would be a lovely cover for anybody, wouldn't it, a place like this? No chance of running into people you don't want to see. All these members here are far more interested in preserving their own little bit of privacy. They've no interest in poking their noses into other people's business. Be a good place for a quiet chat."

Varden came back at that moment and interrupted their thoughts.

"Yes indeed. Mr Latimer paid us a visit last evening. It was early on, before we got too busy. That other person too, Parks, he was here. They didn't stay very long, but then, they never do."

Hudson felt like patting him on the head.

"Any idea what time, sir?"

The 'sir' was not lost on Mr Varden, who looked gratified.

"We don't ask people to enter a time, but it would

have been about nine o'clock. No later than ten, in any case."

So Latimer and Georgie Parks were in the Arena two or three hours before Parks was murdered.

"You've been extremely helpful, Mr Varden, and I'm most grateful for this information. We may have to trouble you for a more formal statement, but I'll try to see that neither you nor the club is unduly embarrassed."

If his last remark had been intended to mollify Mr Varden, it wasn't very successful. The man's face was ashen.

"Formal statement?" he echoed weakly. "Is this absolutely necessary?"

"I'm afraid so. You see, someone else shared your dislike of Mr Parks, but they took it a stage further. You are one of the last people to have seen him alive. A few hours after he left here he was murdered."

"Murdered. Murdered," repeated Varden feebly.

"Please don't get up. We'll see ourselves out."

Eleven

Blackie Webber opened one eye and closed it quickly
as the uncompromising glare of the hospital lights
seared in. The worst of the pain was gone now,
soothed by the deft attentions of the emergency staff.
His mind, no longer distracted by the pain areas in his
battered body, was free to concentrate on the position
he was in. He could almost wish it wasn't, because
there was no doubt about it, he was in schtuck.

Partly his own fault. There was no doubt about that
either, and it didn't help him to feel any better. The
jewellery dodge was not his game, and he should have
left it alone. Greed, that's what that was. A chance to
pick up a couple of quick grand, no questions asked,
and he'd walked into it. Mind you, be fair, he wasn't
to know that Posh Peter and that lot were out looking
for him. Still, he should have had his eyes open a bit
more. With Georgie Parks dead, and him having kept
quiet for a bit, he ought to have expected somebody to
get a bit saucy. There was always some clever bastard
who thought it was time to move up in the world.

Caught him a treat, too. He'd only just turned off a brilliantly lit main road into a side-street when the car pulled in beside him. Posh Peter and three of his clowns jumped out, and lashed into him before he knew what was happening.

"Evening, Blackie."

That's what he'd said, the saucy sod. Well, he hadn't done a good enough job. They'd hardly got started before the patrolling squad-car spotted what was going on and came to sort it out. God bless our police force, that's all. Posh Peter would have put him out of business for life, given another couple of minutes. As it was, he'd be up and around in two or three days, and then we'd give him a 'good evening, Blackie'. 'Goodnight, Peter', more like, and the sooner the better.

Mind you, he'd have to worm his way out of this jewellery lark, first. That might not be too easy. A lot depended on Chief Inspector Actor Bleeding Hudson, and what he wanted out of all this. It would be too humiliating for words for a man of Blackie Webber's standing to go down for possession of stolen goods. Wouldn't do at all. As things stood, look at it from Hudson's point of view. What had he got? Some nicked gear in Blackie's pocket. There was no way he was going to be able to stick Blackie with the actual job, whatever it was. Blackie was genuinely innocent of that, and wouldn't have too much trouble proving it. His life had been blameless for quite a while now, and he thanked his lucky stars for that streak of caution in his nature, which made him keep a careful record of his doings against just such an eventuality as this. No. Actor Hudson couldn't swing the burglary, or the smash and grab, or whatever, on him. All he had was possession.

There could be a trading situation here. The only man who could connect that stuff with Georgie Parks was himself. What did Hudson want most? A tuppeny ha'penny charge against Blackie, or a lead to the real thieves, and perhaps even the bloke who scragged poor Georgie, rest his soul?

That was the burning question, and he wished he could be more certain of the answer. Good job he'd managed to keep his loaf, up to now. Right from the moment those lovely coppers had rescued him, his innate cunning had taken over. At once he had pretended to be closer to unconsciousness than was really the case. It avoided conversation, and the obvious questions they would be asking. Even as they helped him carefully into the night-emergency unit at the hospital his mind was racing. They would keep him in bed, that was certain. That would mean undressing him, and probably checking his pockets for identification and valuables. A joke, that was. He'd got valuables all right. Several thousand quids' worth of stolen jewellery, and that would make them think. Of course, it wasn't actually marked 'Stolen', but on the other hand, people don't go around with that kind of stuff in their pockets. Not straight people don't. Unless they're merchants in Hatton Garden. No, they'd start checking the missing gear lists straight away, and it wouldn't take very long to identify class stuff like that. Next thing you knew, some senior copper would be standing by his hospital bed giving him all that 'we have reason to believe' patter.

He was going to find himself nicked if he wasn't careful. It was all so bleeding unfair, really. That Posh Peter had a lot to answer for, and answer he would. Well, Blackie Webber was not going to have his collar felt by the locals. That would be too much altogether. Just too humiliating that would be. As they helped him on to the bed, he croaked, "Hudson. Get Actor Hudson."

"He's trying to say something," said a welcome voice. "What's that, me old son?"

"Hudson," muttered Blackie feebly, "Scotland Yard. Serious Crime."

The younger police officer shook his head.

"Don't get it," he murmured. "Something about the Yard, Serious Crime, I think he said."

His older companion considered this.

"Couldn't mean the Serious Crime Squad, I s'pose? Never know, round here. I'll bleep the station. Can't do any harm."

Blackie relaxed. All he had to do now was wait.

Hours went by. They'd left him alone at last, and he didn't have to make such a pretence about his condition. Not that it wasn't painful. Especially when he tried to move his – Gawd, mustn't try to do that again.

Footsteps at last, and the sound of a turning door-handle. Keeping his eyes almost shut, Blackie waited for images to move into the narrow crack of vision he had permitted himself. Mr Actor Hudson was grinning down at him. Fat lot of sympathy there, rotten sod.

"Well, well, me old Blackie. Fell off our bike, did we? This had better be good, son. Dragging me down here in the middle of the night. And what's the idea, telling the local talent you work for me? Naughty, that was."

"Is – is that you, Mr Hudson?" croaked the man in the bed, weakly.

"Large as life," agreed the visitor cheerfully. "What's this all about? And stop buggering about with the dying lark. I know, Blackie. Constitution like a cart-horse. Needn't worry, there's no witnesses."

No feelings. Typical bleeding copper. Cautiously, Blackie opened one eye. It was quite true. There was no one else present.

"Wanted to mark your card, Mr Hudson. Do you a favour, like."

"Oh yes? That's nice."

Sam Hudson had a presentiment about this new development. Like any experienced investigating officer, he often reached a point in the progress of a case where some sixth sense came to his aid. A feeling that he was on the right course, and that pieces were about to start fitting together. He had been interested

at once when told that Webber had been attacked and
was in hospital. But the interest had been of a routine
nature. When he learned that the injured man was in
possession of jewellery which had been stolen during
the very burglary Mr Walters had spoken of earlier
that night, his interest trebled at once. Could it be
coming together? The string of burglaries, the Georgie
Parks murder, the missing Mr Latimer? And not
because of some brilliant piece of deduction on his
own part, but because some tearaways had felt like
having a go at Blackie Webber. Ah well, it was the end
product that mattered.

"What's this favour then?" he demanded.

Blackie grinned from a bruised mouth, wincing at
the same time.

"Am I in any trouble?"

"Trouble?" Hudson's voice was heavy with
sarcasm. "I don't know about trouble. You're in
possession, after all. That stuff ties in with several
other cases. You could get ten years, if you call that
trouble."

The man in bed shook his head.

"No way. You know me. Careful. I didn't do them
things. Not my style, and you know it. Nothing to
connect me, Mr Hudson. You'd never make it stick.
Besides, you don't want me. Not on a silly little charge
like this. I'd be out in six months. You'd look daft. No.
What you want is some information I can give you.
About where I found this stuff – "

"Found it?" was the incredulous interruption.

"Found it," he repeated. "That takes you a long way towards many things. Like the real tea-leaves, for a start. And perhaps even somebody who done a murder. Oh yes, I can see that interests you, Mr Hudson. Bit more to chew on than a tuppeny ha'penny 'caught in possession' charge, eh?"

"Are you offering to do me a swap?"

"That would depend," hedged Blackie.

"On what?"

"On whether my lawyer can be present, as a kind of witness. Not that I don't trust you, Mr Hudson, you know that. More like a sort of precaution, you might say."

"Cheeky sod," Hudson glowered down at him.

"Well? What about it?"

"I make no promises, mind. But I'll do what I can. What's his telephone number, this upholder of justice of yours?"

Blackie relaxed. One thing about this Hudson, he kept his word.

It was going to be all right.

* * *

Doctor Keith Firbank was beginning to lose his patience. It was irritating enough to be brought into the police station without any breakfast, but to be kept waiting for almost half an hour was really too bad.

His splendid night's sleep, so thoroughly planned and richly deserved, had all gone to pieces, for one thing. That had been Nancy Parfitt's fault. Coming to his house with such blatantly obvious intentions, and getting him all stirred up, then pushing off without a word. It was enough to unsettle any man. Although, wait a minute, that wasn't quite fair. She wouldn't have gone like that if it hadn't been for those blasted Arena people. They were the real culprits, sending him all that tripe through the post. Well, they may not have heard the last of it. He'd take legal advice about that. Wasn't there some offence involved? Misuse of Her Majesty's postal service, perhaps? Invasion of privacy? There must be something.

Whatever the cause, he'd been too upset to go to bed. As a result, he'd drunk far more than he intended and spent a fitful night. Then, this morning, the police had arrived, just as he was getting his muddled thoughts together for the coming day. Been a bit pushy, too, wouldn't even let a man have his breakfast first. Rather urgent, sir. That's what they'd said. Rather urgent. Made it clear that he was wanted immediately, and not as friendly and polite as he was accustomed to. Well, if it was all that important, he wasn't going to make a fuss.

But that had been half an hour ago. Plenty of time for him to have had his breakfast, set himself up for the day. Instead of which, here he was, cooling his heels in this draughty corridor, with only the duty sergeant on view, and that worthy made it clear he was too busy for conversation.

"Look here, sergeant, are you sure Superintendent

Brent is actually here?''

The sergeant looked up from the desk. He was just as puzzled as the doctor, but he had his orders.

"Quite sure, sir. Won't be long now, I expect."

"Well, couldn't you just ring through? Remind him I'm here?"

The officer looked pained.

"Oh dear me no, sir. That wouldn't do at all. The superintendent will see you as soon as he can."

In his office Michael Brent looked at his watch. We'll just let the good doctor stew for another five minutes, he reflected. It was a very old investigator's trick, but none the less efficient for that. Take one suspected person, let them know they are to be questioned, and leave to simmer in an unfriendly atmosphere. Even the calmest man could not prevent his thoughts from racing around, his imagination from working. It always unsettled people, the waiting, the not knowing, the anticipation of the unexpected.

Brent was looking forward to his session with Keith Firbank. The world was full of surprises, even for a seasoned copper like himself. In any betting situation, he would have taken long odds against Firbank being other than a completely normal man. Just goes to show, you never can tell. Reading between the lines, he was fairly certain that the blondie had been just as surprised as he was. Dressed as she had been, she hadn't gone to the doctor's house for any file, no matter what she claimed. Did she think he'd come up the Clyde on a bicycle? Wonder she hadn't said she wanted to borrow a cup of sugar. No, no. Mrs Parfitt had gone to that house to get herself – watch your language, my son. For indelicate purposes, let's say. That's good enough, covers a multitude of sins. Oh yes, very neat. Anyway, that's what she'd gone for, and finding out what he was had blown her mind. She must have been out of control when she drove off, and P.C. Hibbert was in the local hospital as a result.

The girl wasn't really to blame for that. She'd only been the person at the wheel. The real culprit was sitting outside, and he was going to find that people who caused injuries to police officers were on a very dicey wicket. The murder of Georgie Parks was one thing, and the lawyers would deal with that. The injuries to Tom Hibbert were a different kettle of fish, as Doctor Queer Bleeding Firbank would discover. It was lucky for him the damage was only superficial, otherwise he might have found himself resisting arrest, which could be a painful experience.

D.S. Joe Grayson watched his chief's face, and kept very quiet. This was a side of Barker Brent he'd only previously heard about, and never seen close up. Doctor Firbank had better watch himself at the coming interview, he decided.

"All right, Curly. Let's have a look at him."

The doctor looked tired, thought Brent. Irritable, too. Good.

"Morning, doctor," he greeted. "Just like your help with one or two points."

Keith Firbank sat down heavily on the indicated chair.

"Look here, superintendent, I've been kept waiting out there for thirty five minutes. This had better be very important. I'm going to be late for my surgery, and I haven't even had any breakfast."

"Oh dear," returned Brent evenly, "I hadn't realised that. Well, perhaps there'll be time later. You'll not have heard about P.C. Hibbert, I expect?"

It was clear from Firbank's face that the question took him completely by surprise.

"Chap who found the body? What about him?"

"He was knocked down last night, sir. By a car without lights."

"Oh dear, I'm sorry to hear that. Is it serious?"

Brent was not going to give a direct answer to that one.

"The doctors are still with him," he said gravely.

Keith Firbank was puzzled. Pity about Hibbert, of course, but what had it to do with him?

"Well, let's hope he'll be all right," he said.

"Yes." Brent sat for long seconds before continuing. "There was a young woman driving the car. She was in a very distressed condition."

"I expect she was. Jolly well ought to be."

He still hadn't tumbled, thought Brent, leaning forwards.

"That's not what I meant, Dr Firbank. She was distressed before she hit the constable. She has made certain allegations. Your name has been mentioned."

Nancy Parfitt, of course. She'd gone off without lights. Allegations? What sort of allegations?

"What's she been saying?" demanded Keith.

"She, sir?"

"Nancy Parfitt, I imagine that's who we're talking about."

"Ah yes. You admit that she was at your house last night then, sir?"

"Don't like the word 'admit'. She was there, yes."

If the silly bitch was claiming attempted rape or something, he could be in serious trouble, thought Firbank.

"Mind telling me what transpired, sir?"

That was another word he didn't like. 'Transpired', indeed.

"Nothing. What's she been saying?"

"We'll get to that, sir. I prefer to hear your side of things."

Firbank gave a detailed report on Nancy's visit, leaving out the sexual overtones. It was impossible to tell from Brent's face whether he believed him or not.

"You say she just rushed out, sir? Why should she do that?"

There was no hesitation before the young doctor replied.

"She'd found something that upset her. Something that was addressed to me."

"Yes?"

Clearly the superintendent was not prepared to leave it at that. Firbank shrugged. What the hell? This situation was too serious-sounding for him to worry about niceties.

"If you must know, it was some printed stuff from a

club in London. A gay club.''

So he was going to brass it out, mused Brent. Dropping his voice, he said confidentially, "Look, doctor, I appreciate this is difficult for you, and I've no wish to cause you unnecessary embarrassment. Would you like Sergeant Grayson to leave the room?"

Grayson was already on his feet, but Firbank waved an arm to stop him.

"No, I'd sooner he stayed. What are you getting at?"

Brent shrugged.

"Well, sir, it doesn't mean such a lot these days, you know, and – "

"Christ," Keith exploded. "Not you, too? I had enough with that silly cow jumping to conclusions last night."

The look of surprise on the superintendent's face was not assumed.

"Could you explain what you mean please, doctor?"

"Too bloody true. Are you listening, sergeant?"

"I am, sir."

Grayson wouldn't have missed this for the world. Firbank nodded.

"Right. Well, get this straight, the pair of you. I am not a homosexual, nor have I ever been. I do not belong to this bloody place, the Arena – "

"But the correspondence, sir?" reasoned Brent.

"The correspondence comes to me because of the mailing-list. That's all. Let me explain. This Arena is a new place. The premises used to be occupied by a very different club, called the Blue Pussycat. I was a member, although I hadn't been round there for some time. A couple of months ago I went there with an acquaintance for an evening out. The place had changed hands, but they had an arrangement whereby members of the former club could spend a guest evening. If they liked the new club, they could join." He laughed bitterly. "They're very discreet, I'll give them that. We were there the best part of an hour before we realised what kind of place it was. Then we left, in rather a hurry. That's all there is to it."

Brent was thinking, nineteen to the dozen, having to reshape a lot of his previous ideas. Almost to himself, he said, "And you had to sign the visitor's book, so that's how they got hold of your address."

"Absolutely. They're going to get a piece of my mind, those people."

The superintendent looked across at Grayson, who kept his features carefully wooden. Doctor Firbank seemed to be driving a bus through his chief's theories, and it was no time for comment.

"You understand, doctor, this is purely routine," said Brent stolidly, "but you mentioned a friend just now. No doubt the gentleman will verify your version of what happened?"

"Hardly a friend," contradicted Firbank. "I met him while he was staying at the Swan for a couple of days, looking for a house down here. It turned out we both belonged to the Blue Pussycat, and we agreed to

Twelve

"You made good time, Sam. There's some coffee on its way."

Chief Inspector Hudson nodded and sat down, sizing up his host. He'd heard about Superintendent Michael Brent more than once over the past few years, and of course they'd spoken on the telephone several times in the last two days. The man's physical presence was commanding and powerful. He was the sort of man whose side you were glad to be on, because he would make a formidable opponent.

"I could use a cuppa," he conceded. "My mealtimes seem to have got rather erratic lately. Still, all in a good cause, eh? After this little lot I reckon we'll be entitled to a large steak."

Brent nodded. He, in turn, had been taking his visitor's measure, and liked what he saw. His judgement, as he well knew, was biased heavily in the Yard man's favour because of the unstinted co-operation which had flowed between them.

"Well, we seem to have been lucky, don't we? Your end and my end fit together ver well."

"They do," agreed Hudson, "but I'm not too keen on the lucky aspect. Good, old-fashioned coppering, that's what I prefer to call it. Plus the fact that we didn't get all clever with each other. If you and I hadn't hit it off on the blower this lot could have dragged on for weeks. Ah."

He stopped, as a young W.P.C. came into the room with coffee and biscuits.

"Cups and saucers," he commented admiringly. "You live well down here."

"Only for important visitors," demurred Brent. "Usually it's brown mugs."

The girl went away, and the superintendent stirred at his coffee.

"Who's going to start?" he queried.

"I will, if you like. It's your murder, but it seems to be my villains behind it."

"Please do. I want to hear about this man Latimer."

"Ah yes," sighed Hudson, "an interesting man, my Mr Latimer. Sorry, I should say your Mr Latimer."

Brent smiled expansively.

"Our Mr Latimer, I think."

"Fair enough. A clever chap, and he worked out a good scheme. This security system firm he works for is beyond reproach. Absolutely. And so they have the confidence of most of the county people. Mr Latimer is one of the senior executives, and as such he has been able to pick up quite a few pointers on police procedure. He knows, especially, how each force guards its autonomy, and that was the first thing he used in his own little plan. It was his aim not to use the same constabulary twice, and he achieved it. You see, in the ordinary way of business he goes into premises, studies the lay-out, makes recommendations for a burglar alarm system, and quotes a price. If the customer opts for it, fine. In goes the system, and that's the end of it."

The superintendent was listening intently.

"I see," he interjected. "But not every customer will accept the quotation. Those that don't are put to one side. Latimer has full details of the premises, weak points and so forth. All he has to do is arrange a visit in his own good time. Neat."

Hudson nodded his assent.

"Very neat," he agreed. "And he had the good sense to be patient. There was a gap of a year or more between his quotation and the burglary. On top of that, he didn't even do the job himself. He looked around for a good, reliable villain to carry out the actual burglary, while he set himself up no doubt with a cast-iron alibi each time. Not that there was much chance of his needing it, but he's been a very cautious man. We'd have got to him in the end, naturally, but the process was certainly speeded up by this insurance merchant, Walters."

"That's a new name to me," Brent pointed out.

"Yes, of course. Let me explain about him."

Sam Hudson recounted the visit Walters had made to his office, and the discussion they had.

"So you can see, I was very interested indeed in having an early chat with Hugo Latimer, and it was disappointing to find he'd flown the coop. That was all off my own file, of course, the series of burglaries. Then, when Blackie Webber got sorted out, and he was carrying jewellery missing from the last job, I suddenly found there really was a strong tie-in with the murder of Georgie Parks. I naturally phoned you straight away."

"And I was very pleased you did," Brent replied. "I'd never heard of Latimer, but I had some nice circumstantial building up in the direction of one of our young G.Ps. Chap named Firbank. It seemed he was one of the Arena crowd, and Georgie Parks had their matches in his pocket. For all I knew, Firbank might have been leading some kind of double life, calling himself Latimer while he was in London. Of course, I didn't know all the insurance side of it, then."

"No," Hudson assented. "What about this doctor, then? Did you already know he was bent?"

Brent shook his head in denial.

"That's just it. The poor sod is straight as a die. The Arena just got hold of his address more or less by mistake. Unfortunately for the doctor, some young woman found some papers in his house, and jumped to the obvious conclusion. Spoiled his evening altogether, if you take my meaning."

Hudson chuckled.

"That's what I call real bad luck. Especially if she tells all her friends."

There was no answering smile on Brent's face.

"That young woman would be very wise to keep her mouth shut, and I shall be telling her so before the day's out. However, we've more pressing things to do. Our Mr Latimer must come first. Any idea why he should have killed Parks?"

"I've got a theory about that. A man with a highly respectable cover like Latimer has to protect it. The fact that he used to meet Parks at the Arena suggests that he wanted to keep their relationship on neutral ground. Didn't want Parks encroaching into his ordinary daytime life. It's my guess that Georgie wasn't satisfied with knowing so little about him. You know how the villains hate mystery men. I think we might find that Parks had traced him to the Shepherd's Bush address, and was putting a bit of pressure on him. More jobs, a bigger share, who knows? But whatever it was, it was making life uncomfortable for Latimer. He decided to move out, quit London altogether, come down here to Cookstone. Throw Georgie off the scent altogether."

"But Parks was too good for him, and traced him here anyway," mused Brent. "Yes. I can see the sense of that. Latimer could have felt he was being pushed beyond endurance, and decided that killing Parks was the only answer.''

"Something like that. And with almost no risk, you see. There was nothing to tie him in with a villain like that. He had a very good chance of getting away with it. I think he's going to be a bit surprised when we turn up.''

"I hope so. Anyway, you'll have him on these burglaries, won't you? That ought to be watertight.''

Hudson raised his eyes upwards.

"I do hope you're right. One way or another, if a couple of old coppers like us can't screw him for something it'll be a bloody fine look-out.''

They grinned at each other, with mutual respect.

It occurred to Brent that the question of Parks lady-friend seemed to have been put to one side.

"What about the girl who went missing? What was her name – Marshall? Where do you think she fits in?''

"I'm not sure,'' muttered Hudson, "but it's my guess she comes into the silly bitch category. You know, what's a nice girl like you doing with a character like this? That sort of thing. I'll have her, though, when she comes back. Frighten the bloody daylights out of her. She'll be a bit more careful with her playmates by the time I'm finished.''

Mike Brent looked at his watch. "Right then, I'll just see if our warrant is ready yet.''

He picked up the telephone and spoke into it. Detective Sergeant Grayson came in on the double, bearing the official document. The superintendent examined it closely.

"First class," he grunted. "Well then, that about ties it up. If you're all set, chief inspector, we can be on our way."

Hudson rose, smoothing at his jacket.

"Right beside you, superintendent."

Ten minutes later a plain blue car turned into an open gateway and came to a halt in front of an open oak door.

"Silly, that is," commented Brent. "We could be common prowlers, for all this chap knows."

"I agree," said Hudson. "Or even burglars. You never know these days. And what are the police doing about it all? That's what I'd like to know."

They climbed from the car and walked across crunching gravel to the door, with Sergeant Grayson bringing up the rear. Superintendent Brent pulled at the old-fashioned bell, and there was a noisy clanking inside.

They heard footsteps, and a man came to the door. He was about thirty years old, with an open, trustworthy face. The jury were going to love him, reflected Brent sourly.

"Mr Hugo Latimer?" he enquired pleasantly.

"Yes, that's me. What can I do for you?"

"We are police officers. We have reason to believe you are in a position to give us certain assistance with our enquiries."

Latimer started to speak, changed his mind, turned

abruptly and walked back inside.

The senior officers nodded to each other, and followed him in.

Sergeant Grayson stood on the porch, feeling in his pocket. There'd be plenty of time for a cigarette.